MW00717392

Boris A. Krylov

COLD WEATHER CONCRETING

CRC Press
Boca Raton Boston New York Washington, D.C. London

New Directions in Civil Engineering

SERIES EDITOR: W. F. CHEN *Purdue University*

Response Spectrum Method in Seismic Analysis and Design of Structures
Ajaya Kumar Gupta *North Carolina State University*

Stability Design of Steel Frames
W. F. Chen *Purdue University* and E. M. Lui *Syracuse University*

Concrete Buildings: Analysis for Safe Construction
W. F. Chen *Purdue University* and K. H. Mossallam *Ministry of Interior, Saudi Arabia*

Stability and Ductility of Steel Structures under Cyclic Loading
Yuhshi Fukumoto *Osaka University* and George C. Lee *State University of New York at Buffalo*

Unified Theory of Reinforced Concrete
Thomas T. C. Hsu *University of Houston*

Flexural-Torsional Buckling of Structures
N. S. Trahair *University of Sydney*

Advanced Analysis of Steel Frames: Theory, Software, and Applications
W. F. Chen *Purdue University* and S. Toma *Hokkaigakuen University*

Water Treatment Processes: Simple Options
S. Vigneswaran *University of Technology, Sydney* and C. Visvanathan *Asian Institute of Technology*

Simulation-Based Reliability Assessment for Structural Engineers
Pavel Marek *San José State University,* Milan Guštar *Prague* and Thalia Anagnos *San José State University*

Analysis and Software of Cylindrical Members
W. F. Chen *Purdue University* and S. Toma *Hokkaigakuen University*

Fracture Mechanics of Concrete
Zdeněk P. Bažant *Northwestern University* and Jaime Planas *Technical University, Madrid*

Artificial Intelligence and Expert Systems for Engineers
C.S. Krishnamoorthy *Indian Institute of Technology* and S. Rajeev *Indian Institute of Technology*

Introduction to Environmental Geotechnology
Hsai-Yang Fang *Lehigh University*

Fracture Processes of Concrete
Jan G.M. van Mier *Delft University of Technology*

Cold Weather Concreting
B.A. Krylov *Design and Technology Institute (NIIZhB), Moscow*

The Finite Strip Method
Y.K. Cheung and L.G. Tham *University of Hong Kong*

Acquiring Editor: Navin Sullivan
Project Editor: Albert W. Starkweather, Jr.
Cover design: Dawn Boyd

Library of Congress Cataloging-in-Publication Data

Krylov, B. A.
Cold weather concreting / Boris A. Krylov.
p. cm.
Includes bibliographical references and index.
ISBN 0-8493-8247-4 (alk. paper)
I. Title.
TA682.43K79 1997
624.1′834—dc21

97-33292
CIP

International Standard Book Number 0-8493-8247-4
Library of Congress Card Number 97-33292
Printed in the United States of America 1 2 3 4 5 6 7 8 9 0
Printed on acid-free paper

The Author

Boris Alexandrovich Krylov, born in 1926, is a doctor of engineering, professor, and academician of the Academy of Architectural and Building Sciences of Russia, of the Russian Engineering Academy, and of the International Academy of Informatization. Dr. Krylov is an expert in the field of construction under extreme conditions, including Arctic and permafrost regions and regions with dry, hot climate, and in problems relating to acceleration of concrete hardening rate at precast concrete factories and at building sites. He is the author of more than 300 publications, inventions and patents. Dr. Krylov represented the U.S.S.R. at the Reunion Internationale des laboratoires d'essais et de recherches sur les materiaux et les constructions (RILEM) for many years, heading the Technical Committee on Winter Concreting (39-BH) in this organization. He presently is Deputy Director for Research of the Research Institute of Concrete and Reinforced Concrete in Moscow and a professor at Moscow State University of Construction.

Introduction

Concrete is a basic material in modern construction. It is used in practically all building or engineering structures erected today and will continue to be in the forefront of building materials for decades to come. Many future generations will have to deal with it and components made of it. Seemingly simple, concrete still remains the subject of intensive studies conducted by research institutions and laboratories in all industrialized countries. Every time there is a change in the environment and service conditions of structural members made of plain or reinforced concrete, the material has to be retested because it is very sensitive to such changes.

Many varieties of concrete have been developed for applications in structures capable of reliable and long service when subjected to various degrees of heat and cold, humidity, radiation, or corrosive environment. Any structure can be made of concrete but its broad engineering potential not always agrees with economics. This is the reason we always have to justify a solution economically and try to find the most efficient uses of the material. There are, of course, traditional applications of concrete, where it holds a monopoly, such as foundations, hydraulic dams, airfield runways, large water ducts and headers, cooling towers, and many other structures where any substitution of concrete is improbable today. So concrete is used in construction in enormous quantities. A total of 130 million cu.m of precast concrete components was produced annually in Russia and former Soviet Republics alone and about the same volume was cast in place.

The thorough development of regions with severe climate for the last few decades always involved construction projects. There builders encountered considerable difficulties when they tried to use concrete. The specific climatic conditions of Arctic or Antarctic regions impose certain limitations on concrete building construction and maintenance of reinforced concrete structures. As mentioned above, we can use concrete in erecting any structure in any environment, including harsh climate, but, in addition to high quality and durability, we need to minimize costs and labor input. So it is not only a technical problem but also a problem of cost effectiveness. To solve it, one should know concrete itself quite well and, moreover, methods of concreting in cold weather, of curing until concrete attains its design strength, and other characteristics of the material.

The purpose of this book is to acquaint specialists with specific features of concrete construction in cold weather, with effects of frost on the structure and properties of concrete, with methods to be used to have concrete harden when the ambient temperature is below 0°C, and with other winter concreting problems.

Contents

1 Winter Construction

1.1 SPECIFIC FEATURES OF WINTER

Winter has some peculiarities that affect construction in general and concreting in particular. Its duration is different in different parts of the globe, but cold weather with light frosts also may happen in spring and autumn — not just in winter. So the cold season is longer than winter according to the calendar.

Thus, in the central part of Russia, the cold period, including winter, early spring, and late autumn, may be as long as five or six months reaching eight to 10 months in the north. The situation is the same in Canada, Alaska, northern China and Japan, Finland, Norway, and Sweden.

Winter brings subzero temperature of the outside air, which may drop to –50°C (to –84°C in the Antarctic), and, besides, high wind, sometimes gales and heavy snowfalls with snowstorms that may last for days. The snow cover may be several meters thick. People in Arctic villages make manholes in their roofs in addition to usual doors in their houses to be able to crawl outside after a heavy snowfall that sometimes may cover the house up to the roof. It has happened in northern Siberia that strong, long snowstorms buried power transmission lines on wooden poles six to eight m high completely in the snow. So the poles had to be searched in snow drifts and dug out.

Northern regions in Russia, Canada, and Alaska are covered by a deep layer of permafrost. The temperature of the frozen ground below the active layer* is stable and ranges from –1°C to –3°C. Permafrost often contains a lot of moisture (up to 30%) and turns into pulp when it thaws. The bearing capacity of this ground is very low and builders in the permafrost regions prefer to keep the ground frozen not only during construction but also after a building is completed. For permafrost construction projects, concreting must be of winter type practically the whole year round.

1.2 CONCRETING IN WINTER CONDITIONS

Concreting in winter conditions is quite difficult. The reason is that concrete can harden and develop high strength in a relatively short time only within a certain range of above-zero temperatures. It is possible to create favorable conditions for concrete to harden when the ambient air temperature is below zero but that requires additional energy, materials, and labor. This is the reason why builders in many European countries prefer to minimize or stop construction altogether on cold days.

* Active layer of the ground is a layer down to 6 m thick, which thaws in summer and freezes in winter. Its temperature at the surface is close to that of the ambient air.

However, in other countries situated in the moderate or northern climatic zones long stoppages of construction in winter, concreting included, are uneconomical due to long downtimes of equipment and workers. It is better to bear additional costs trying to minimize them as much as possible.

An analysis has shown that reinforced concrete structures cast in place in winter cost 20 to 30% more than in summer. The cost is brought up mainly by expenditures required to make concrete develop its critical* or design strength. According to Russian regulations, the critical strength of concrete depends on its class. It is 50% for classes up to B4, 40% for classes up to B25 and 30% for classes over B25. But if a reinforced concrete structure must meet higher specifications for frost resistance (e.g., hydraulic structures subjected in operation to freezing and thawing cycles in a water saturated condition), its strength should be at least 80% by the time it freezes.

Concrete frozen after it attained critical strength continues to harden and, as a rule, reaches the strength of the class it was designed for. It is assumed that the structure will not be fully loaded until it develops the design strength.

A question is asked sometimes: Isn't it better to let concrete freeze in any structure only after it developed its design strength? It would be the best solution, of course, but the cost of energy expended to maintain favorable temperature conditions for the hardening of concrete would more than double, which is uneconomical. In some cases, when the construction schedule is very tight and a project is to be completed in winter, the additional costs are inevitable. The simplest and least expensive way is to increase the class of concrete by the required number of steps.

1.3 PROBLEMS OF WINTER CONCRETING

Concreting in winter conditions should be preceded by some preparations and a careful technical and economic analysis. We must solve a number of problems on which the erection time, quality and durability of structures, money and labor expenditures, and the increase in costs as compared with summertime are dependent. They include:

- provision of sufficient energy resources to heat the concrete and shelters for the personnel at the site;
- provision of steam and heat insulating materials and of instrumentation needed to cure the concrete and to monitor its hardening;
- solution of problems involved in concrete mixing, particularly if the mixing plant is at the construction site;
- coordination of concreting operations, including above all transportation and placing of mixed concrete, particularly in the frost or under a snowfall;
- creation of favorable conditions for the hardening of concrete, identification of the best methods of curing in the frost after placing, and preparation of the necessary equipment in advance;
- provision of comfort for the workers as much as possible.

* Critical strength is the level of compressive strength attained by concrete by the time it freezes, after which freezing does not affect much its properties after thawing and during further hardening.

When these problems are resolved well in advance, there is every reason to believe that the concrete structure will be erected competently in winter at a minimal cost whatever the ambient temperature.

2 Effects of Frost on the Structure and Properties of Concrete

The subzero temperatures characteristic of the winter season greatly affect the structure and properties of concrete. The frost is especially harmful to green concrete that still contains a lot of mechanically bound (free) water while its structure is not yet formed enough to resist destructive factors. As the concrete hardens, the damaging effects of the frost diminish and the freezing becomes harmless for a structural member when it reaches a certain level of maturity.

It should be noted, however, that many freezing and thawing cycles in a water saturated condition result in deterioration of concrete even if it developed its design strength. It does not mean that concrete and concrete structures are not durable. They may stand for centuries provided the concrete composition was selected correctly taking the environment into account. Remember that even granite may gradually deteriorate when subjected to alternating cycles of freezing and thawing. So there is no reason to be afraid of building from concrete in any climatic zone and in any season. The proof is the innumerable number of buildings and engineering structures built of concrete in many countries. They have been in use for years and years and are in good shape. Their life is hard to define since it is rather obsolescence than physical wear that may put them out of service.

This book does not discuss the effects of frost on hardened concrete since this subject has been studied by many researchers and is well described in literature. But the effects of frost on concrete at an early age are very important for construction in winter, they were less studied, and need proper attention.

2.1 MODERN VIEWS ON HARDENING OF CONCRETE

The hardening of concrete, whereby a plastic mass turns into an artificial stone that has good physico-mechanical characteristics, is accompanied by a set of very complex phenomena that have not been studied well enough yet and are not fully controllable. The reasons are the versatility of active ingredients of the original binder (cement), the complexity of the system of new formations, and the multiplicity of components in the material as a whole, which include substances in different states of aggregation. Due to relaxation with time, the microstructure and macrostructure of concrete varies and it becomes a new system with different properties at each hardening step.

The processes of chemical and physical transformations slow down considerably as temperature drops and go faster as it rises. So the temperature factor is regarded as a most effective influence on the hardening of concrete.

The concrete hardening processes were studied by many researchers, including A. A. Baikov, P. P. Budnikov, Yu. M. Butt, V. V. Timashev, Yu. S. Malinin, O. M. Mchedlov-Petrosyan, A. F. Polak, P. A. Rebinder, A. E. Sheikin, G. Green Kenneth, A. Grudemo, Copland, F. M. Li, T. Powers, and others. Their points of view on the hardening differ greatly and cannot be combined into one theory. Thousands of papers were devoted to the concrete hardening problems. They are discussed at international and national conferences all the time and it is impossible to describe them all in this book. I dwell on this issue only briefly and in the most schematic form from the standpoint of further operations with concrete to make it harden under different temperature conditions.

According to the crystallization theory of hardening, which has an overwhelming number of supporters, minerals of the cement clinker are dissolved and interact with water as the cement is mixed with it. The liquid phase is rapidly saturated with calcium hydroxide formed by hydration of C_3S. Gypsum and alkalis contained in the clinker are quickly dissolved.

The saturation of the liquid phase with calcium hydroxide is very fast because it requires only to dissolve CaO that is formed when 0.3 to 1% of C_3S contained in the cement reacted with water. In fact, more of CaO goes into the liquid phase and it forms a supersaturated solution since 1 to 4% of C_3S hydrate within first three minutes after Portland cement is mixed with water even at normal temperature.

Thus, the hydration of minerals of the Portland cement clinker takes place in saturated solutions of $Ca(OH)_2$ rich in caustic alkalis and gypsum. As they hydrate, calcium silicates, C_3S and C_2S, form calcium hydrosilicates and calcium hydroxide. The latter precipitates in the form of prisms that are gradually transformed into lamellar crystals. The calcium hydrosilicates undergo changes that involve transformation into a high or low basic modification depending on concentration of CaO in the solution. Tricalcium aluminate (C_3A) reacts with gypsum to form a high sulfate modification of calcium hydrosulfoaluminate. The stability of the latter depends on the content of calcium hydroaluminate in the solution. It acquires the low basic form when the content is high and produces a continuous series of solid solutions.

Tetracalcium aluminate ferrite (C_4AF) hydrates and some part of it also reacts with gypsum producing the high sulfate form of calcium hydrosulfoaluminate and the iron-containing phase.

Calcium hydrosulfoaluminate decomposes when temperature rises. The process starts at 40°C or 45°C and accelerates sharply at temperatures close to 100°C.

In addition to the minerals described above, Portland cement clinker contains the glass-forming phase that consists of uncrystallized C_3A and C_4AF with inclusions of CaO and SiO_2. When Portland cement is mixed with water, this phase hydrates and forms hydroaluminates, hydroaluminate ferrites, and hydrogranates. The latter crystallize in the form of regular isometric crystals.

A comparison between the hydration rates of monominerals and of their mechanical mixtures shows that the latter hydrate faster than any individual component in its pure form. It is assumed, therefore, that the variation of the hydraulic reactivity of clinker minerals during their joint hydration is the main effect that they produce on each other.

New formations produced by the hydration of cement make up the structure of a new material, i.e., hardened cement paste. The formation of the structure is as complex a process that has not been studied completely as the chemistry of hardening. The mechanism of cement hydration is the central question there and it can be represented on the basis of modern views as follows.

Mixing cement with water triggers dissolution and hydration of clinker minerals. The liquid phase is saturated and, when oversaturation with hydration products is reached, newly formed hydrates begin to crystallize in the form of submicrocrystals. The higher the oversaturation of a solution the faster the crystallization process with the appearance of more disperse particles. Spreading over the whole system, the crystallohydrates form a coagulation structure. As the number of crystals grows, they start to move closer together.

Mechanical admixtures, quartz sand in particular, have a favorable impact at this stage of hardening. Acting as an active backing, they help to accelerate the crystallization, to dissolve the initial binder, and to reduce the probability that the nuclei of a new phase will be formed on the surface of the latter's grains. Thus, the presence of the quartz aggregate should weaken the isolation of cement grains and increase the degree of hydration postponing the retardation or discontinuance of the hydration until later.

The acceleration of crystallization and hardening at the boundary with the aggregate is accompanied by an increase in the strength of interface layers of the hardened paste.

Strong crystallizing contacts of intergrowth are formed between individual small crystals of hydrates in the process of crystallization. Not all crystal hydrates in the hydrated cement paste, however, can form regular intergrowths. The priority here is given to crystal hydrates of the same structural type. It appears that intergrowths of hydrosilicates and hydroaluminates lead an independent existence in the mass of the hardened cement paste. Besides, the structure carries a great number of individual hydrate crystals and their aggregations whose individual particles intertwine and are held together by mechanical forces of adhesion.

Intergrowth contacts may arise only after many particles of new formations appear in the system and their spacing does not exceed the double thickness of the adsorption layer of hydrate molecules. The crystals join into individual aggregations who intergrow to form the crystal skeleton of the hardened cement paste, a solid artificial polymineral. Then new crystals grow around the skeleton filling its voids and reinforcing it. Gradually a system with a weak coagulation structure turns into a crystalline structure. The amount of gel decreases because the particles grow and join each other and also because water is sucked to the nuclei of the cement particles which have not hydrated yet. The system becomes stronger and more rigid.

The spontaneous crystallization of hydrate forms gives rise to thermodynamically unbalanced contacts that distort the crystal lattice of the material. So later, when concrete is cured in wet conditions, there is recrystallization that starts by itself and results in dissolution of thermodynamically unstable compounds and in growth of regularly formed crystals.

The recrystallization weakens the structures at certain stages and reduces the strength of the hardened cement paste. So the strength of the hardened paste depends on the growth of small crystals, their joining into a rigid skeleton, and the reinforcement of the skeleton by new small crystals that grow on it. Any recrystallization in an intergrowth already formed can only weaken its structure.

It should be noted that the growth of the crystal skeleton not only increases strength but also causes internal stresses due to directed growth of crystals that are already bound by intergrowth contacts. The internal stresses (similar to the crystallization pressure) weaken the system because they partially destroy it in the weakest sections. The strength of Portland cement usually increases in hardening because the destructive process is more than compensated by the constructive process of formation of new intergrowth contacts and an increase in crystal sizes.

An increase in temperature does not affect the quality of the hardening mechanism of the Portland cement binder but increases the rate of the structuring processes (accelerated dissolution of the initial material, fast growth of smaller crystals in new formations, etc.). Imposition of vibration effects at the stage of the coagulation structure helps to bring together and to pack more closely the particles and to form close coagulation bonds necessary to form the crystalline structure. According to A. F. Polak, a close coagulation structure is 10 to 100 times stronger than that with a widely spaced coagulation, i.e., when the distance between particles exceeds the double thickness of the adsorption layer that consists of hydrate molecules.

The formation of the hardened cement paste microstructure depends to a large extent on what happens to water at the interface of the cement grains. Hydration of the cement grains involves bringing a solvent to the surface of a grain, chemical reactions on the surface, and removal of the reaction products from the surface. These steps are consecutive and the rate of the process depends on the slowest one, which is the transfer of matter or the internal mass transfer, since isolating films of new formations on the surface of cement grains will prevent moisture from getting to the unhydrated part of the cement grain.

Thin films of hydrosulfoaluminates and calcium hydrosilicates formed on the surface of hydrating cement grains at the start of the process can be easily penetrated by water that diffuses from the intergrain space and from the saturated liquid phase which enters the intergrain space. As the films thicken and condense, the diffusion of ions through it becomes more difficult with time. The liquid phase in the space between the colloidal film and the cement grain surface (transition zone) becomes more difficult to penetrate and more saturated as compared with the liquid phase in the capillary space. The difference in concentrations causes osmotic pressure.

The gel-like films dissolve by themselves after a time and are destroyed by osmotic forces and by the increase in volume of new formations that crystallize on the internal surfaces of the films. So the reaction of the cement with water is accelerated again. When the temperature is high, films formed around cement grains at the early stage of hardening deteriorate very quickly and the unreacted parts of the grains become exposed again to moisture. This phenomenon agrees well with the nature of heat evolution that has extremums at early stages, which correspond to the initial period of hydration and to its new increase after the films are destroyed.

As the hardening progresses, the hydration processes slow down considerably and become very weak after 28 days of curing in favorable heat and moisture conditions. Strength properties of concrete may decrease and increase again from time to time as a result of the recrystallization processes and accumulated internal stresses.

As was mentioned above, an increase in temperature affects greatly the hardening process and formation of the microstructure and macrostructure of concrete. Lower viscosity and higher reactivity of water due to heating help to accelerate the dissolution of the binder, to oversaturate the liquid phase, and to produce a great number of crystallization nuclei and a finer dispersed structure. This in turn reduces internal stresses under optimum conditions of thermal treatment.

An accelerated temperature rise that may be produced by electric contact heating allows to destroy the unstable films around hydrating cement grains and to achieve practically the same degree of hydration as in mild conditions of thermal treatment.

A higher hardening temperature helps to turn unstable compounds into more stable ones (e.g., hexagonal crystals of hydroaluminates into crystals of hexahydrate tricalcium aluminate and high basic calcium hydrosilicates into lower basic ones). So the thermal treatment can shorten the time of physico-chemical processes in concrete as compared with hardening under normal conditions.

The strength characteristics of concrete are known to depend not only on the strength of the hardened cement paste structure but also on the structure of the system as a whole and on reliability of bonds between components under external loading. From this point of view, the structural and physical phenomena observed in concrete under thermal treatment are quite important.

Heating of concrete as of any physical body results in expansion of its constituents that have different thermal properties. The expansion coefficients of the concrete's solid components do not differ much but those of water and entrapped air differ from the expansion coefficients of the solid components by one and two orders of magnitude, respectively. A temperature of 60°C or higher accelerates internal evaporation of moisture in the capillary-pore system (i.e., concrete), which increases the content of the vapor-air mixture in the material and thus its internal pressure, loosens the concrete, and decreases its density and strength.

Despite the favorable role played by the thermal treatment of concrete in acceleration of hardening it may also cause deterioration of the microstructure. The result may be a somewhat lower strength and shorter durability as compared with concrete that hardened under normal conditions. We discuss this subject in detail elsewhere in the book. The adverse effects of the thermal treatment can be reduced dramatically by using an optimum mode of heating or vibration of hot concrete when accelerated heating is applied.

This very short analysis of the concrete structuring process in hardening shows that heating of fresh concrete creates on the whole favorable conditions for accelerated transformation of the material into an artificial conglomerate stone.

The chemistry of hardening remains practically the same under thermal treatment but the reactions go much faster. The temperature factor affects the microstructure of concrete and causes certain defects which may adversely affect the properties of concrete and the durability of structural members made of it. Certain techniques make it possible either to avoid the structural defects or to reduce them substantially.

So the thermal treatment is an effective method for accelerating the hardening of concrete under various temperature conditions of the environment.

2.2 THE PHASE STATE OF WATER IN CONCRETE AT DIFFERENT TEMPERATURES

The phase state of water in concrete has a great theoretical and practical importance. One cannot formulate a competent theory and develop practical methods of winter concreting without studying it in depth. The nature of physico-chemical phenomena in concrete at above-zero and subzero temperatures primarily depends on peculiarities of the water structure and on variation of its properties according to the temperature factor. Solid components of concrete mixture only change their volume with temperature but their structure and ability to react with water remain unchanged.

Each component of concrete, which is a heterophase system, responds to variation of temperature in its own fashion. The gas phase, which consists of entrapped air and of water vapor the air contains, contracts or expands according to Charles' (Gay-Lussac's) Law. Its change in volume is 1/273 for a change in temperature of 1°C.

Solid components, such as cement grains and fine and coarse aggregate granules, become larger or smaller in volume according to the coefficient of linear expansion for a particular substance. This coefficient is 12.6×10^{-6} m/m °C for a cement solution, 0.5×10^{-6} m/m °C for quartz sand, and 8.3×10^{-6} m/m °C for granite. The coefficients of linear expansion of solid components are rather close thus enabling concrete to work in a structural member without any noticeable deterioration under usual temperature conditions (between +50°C and –50°C).

Water,* whose quantity in fresh concrete is quite large (up to 50% or more of the amount of cement), is the only component that behaves differently as an abnormal substance. Water passes all states of aggregation- the solid, liquid, and gaseous states — in a narrow temperature range from 0°C to 100°C. Taking the state of water at 20°C as a reference we can see that its structure is an isotropic system of associations of polar molecules, which contains a negligible number of free molecules of oxygen and hydrogen. The reaction of minerals of the cement clinker with water takes place when a molecular association contacts the surface of a cement grain.

As the water temperature rises, the bond between molecules weakens, large associations begin to disintegrate into smaller ones, they move faster, the number of collisions of the water associations with the surfaces of cement grains in a unit time increases, and the chemical reaction accelerates.

When the temperature rises further approaching 100°C, the bond between molecules weakens so much that the associations practically disappear and the water becomes a system of highly active molecules. Many of them can leave the system freely in the form of steam, visible at a temperature of 60°C or higher, due to the weakening of the intermolecular bonds. The number of contacts of water molecules with the surfaces of cement grains at high temperatures (60°C or higher) increases dramatically and the chemical reactions are accelerated. Concrete can develop a strength equal to 70% R_{28} in seven to 14 days at 20°C, but ,when the temperature is 80°C, it can attain the same strength in six to 10 hours.

* Water in a concrete mixture is a solution of different substances which do not affect its physico-chemical properties, however, except electrical conductivity. For this reason its dissolved substances are not taken into account when the phase state of water in concrete is analyzed.

So higher reactivity of water accelerates the hardening of concrete and this principle underlies all methods of thermal treatment of structural members regardless of energy used for this purpose.

The picture is different when the temperature drops. According to modern views, water as a distinctly polar liquid becomes a system containing a great number of ordered zones that consist of large associations of molecules when its temperature approaches 0°C. It can be regarded as a quasicrystalline system where the internal structure of the liquid is closer to that of crystals than gases but differs in that ordered arrangements involve much smaller volume elements. So water at this stage, although having all attributes of a liquid, should be considered from the standpoint of this state of aggregation.

The reactivity of water decreases due to strengthening of intermolecular bonds as the temperature decreases. And, in addition to the dipole interaction, hydrogen bonds promoting growth of molecular associations also increase. As the latter grow, they become less mobile. A decrease in the thermal motion energy in general due to cooling also helps. All this results in greater viscosity of water, which is nearly 80% higher at 0°C than at 20°C, less collisions of water particles with cement grains, and retardation of reactions between them. It is reflected quite well by changes in setting rates of various cements and by the concrete strength development. The setting time of cement at a temperature close to 0°C is 2 to 10 times longer (according to the type of cement) than at 20°C. Roughly the same relationship can be observed for the concrete hardening rate.

The density of water increases with the temperature drop to a maximum value at 4°C whereupon it decreases again. At temperatures close to 0°C water undergoes structural changes as part of preparations for conversion of the system into the solid state of aggregation.

It is erroneously assumed sometimes that first crystals of ice appear in water in the temperature range between 4 and 0°C. Investigations did not find anything of the kind but the structuring of the water does take place. The arrangement of particles is not chaotic and somewhat resembles liquid crystals due to prevailing distinct orientation of molecules in microvolumes.

The state of aggregation of mechanically bound water starts to change in concrete, a capillary-pore system, at 0°C (at normal pressure), which leads to changes in the molecular structure. This conversion may be instantaneous in a supercooled water when a crystallization center appears. When water changes into the solid state — ice — its volume increases by about 9% because ice is lighter than water. This again characterizes water as an abnormal substance because an overwhelming majority of substances have the highest density in the solid state, a lower density in the liquid state, and a much lower density in the gaseous state.

What is the reason for the lower density of ice as compared with that of water in the liquid state? The explanation can be found in changes at the molecular level.

It is well-known that the water molecule is asymmetric; two hydrogen atoms are connected with an oxygen atom so that their connections form an angle H-O-H of 105°03′ (Figure 2.1). The distance between each hydrogen atom and the oxygen atom O-H is 0.9568 Å and between hydrogen atoms H-H is 1.54 Å.

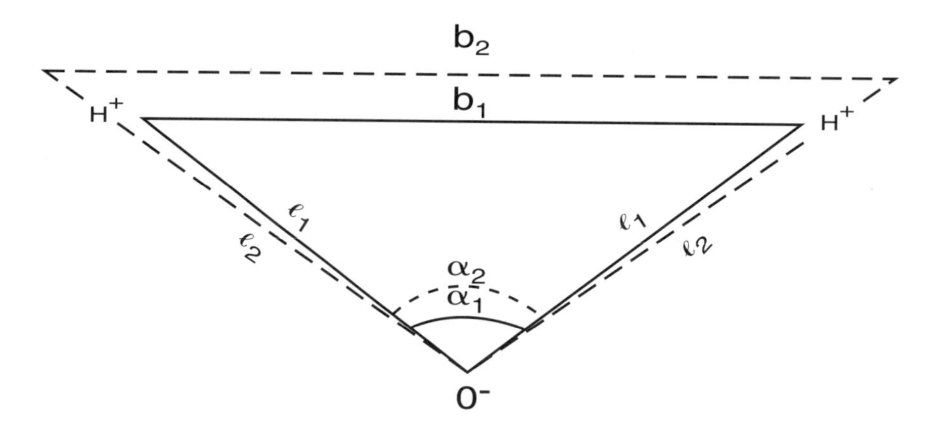

FIGURE 2.1 The scheme of deformation of water molecules as they turn into ice. $l_1 = 0.9568$ Å, $l_2 = 0.99$ Å, $\alpha_1 = 105°03'$, $\alpha_2 = 109°28'$, $b_1 = 1.54$ Å, $b_2 = 1.62$ Å.

As water turns into ice, the structure of the molecule undergoes changes and it occupies a larger volume in space. So the angle H-O-H increases to 109°28′, the distance between the atoms of hydrogen and oxygen O-H to 0.99 Å, and that between the hydrogen atoms H-H to 1.62 Å.

Thus, when water changed to ice, its density decreased due to an increase in the volume of the molecules although the distances between them shortened (the intermolecular spacing diminished from 2.90 to 2.76 Å). The restructuring of water in freezing may produce high pressures up to 2500 kg/cm^2. This is the reason why containers of almost any material (stone, concrete, or steel) completely filled with water and tightly closed break when the water freezes.

Hence, the conclusion: whatever the strength of concrete, it will never be able to withstand without damage the high pressure caused by the freezing of water in it. Its tensile strength is eight to 10 times lower than the compressive strength and naturally its ability to resist tensile stresses is quite low.

Mechanically bound water contained in concrete in capillary pores — macropores with a radius higher than 10^{-5} cm–, in pores of the intergrain space, and between aggregations of gel particles turns into ice at 0°C, as was mentioned above. But the rate of the process depends on the cooling rate of the material. Freezing of water in fresh concrete where its content may be as high as 80% is the most dangerous.

In addition to the mechanically bound water, fresh concrete contains some amount of adsorptionally or physically bound water which changes to ice at lower temperatures. This water is held on the surfaces of mineral particles, which have a surface energy, by electromolecular forces to which water is particularly sensitive because of the polarity of its molecules. The strength of binding water molecules by mineral particles is governed by the surface electromolecular field and depends on the intensity of the field. The latter is a rapidly decreasing function of the distance from the surface of a particle. So the electromolecular forces are enormous very close to the surface (in the order of one or two diameters of the molecules) and

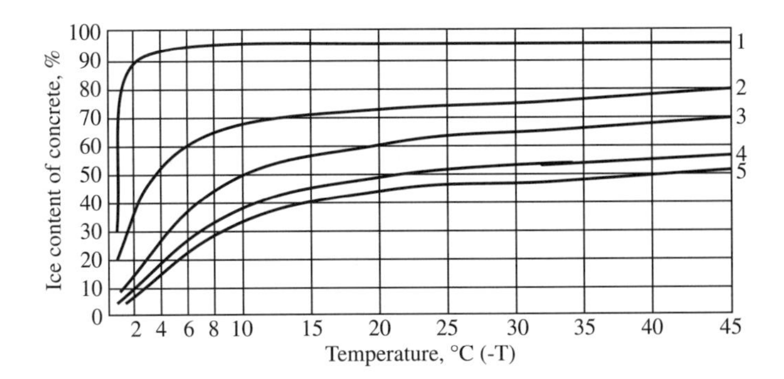

FIGURE 2.2 The phase state of water in concrete of the composition 1 : 2.9 : 4, W/C = 0.72, of different strengths according to the freezing temperature. 1 — concrete frozen after mixing, 2 — concrete frozen when its strength was 15% R_{28}, 3 — concrete frozen when its strength was 50% R_{28}, 4 — concrete frozen when its strength was 70% R_{28}, 5 — concrete frozen when its strength was 100% R_{28}.

diminish fast with the distance. This determines the degree of bonding of water layers at different distances from the surfaces of mineral particles. The most remote layers become solid at a higher subzero temperature than those that lie nearer to the surface. The molecular layers of water on the surfaces of solid particles have gigantic surface tension forces, remain liquid even at –190°C, and are close to solids in their properties.

So physically bound water does not damage concrete much because of its small amount and gradual freezing over a wide range of subzero temperatures. The only exception is mass transfer which will be discussed later.

Chemically bound water, which is part of new formations (calcium hydrosilicates and others), is absent in concrete as such, cannot turn into ice, and responds to a drop of temperature of a new formation by changing its volume according to the coefficient of cubic expansion of a specific substance.

It follows from the above that only mechanically bound water becomes a destructive agent for concrete when it freezes because its freezing point is 0°C and its volume expands (Figure 2.2).

2.3 THE MECHANISM OF DETERIORATION OF CONCRETE DUE TO EARLY FREEZING

The effects of frost on concrete were studied by many researchers, including V. M. Moskvin, I. N. Akhverdov, G. I. Gorchakov, S. A. Mironov, I. A. Kireenko, S. V. Shestoperov, A. A. Shishkin, A. Collins, T. Powers, T. Brownyard, P. Nerenst, and A. Nykenen. At the same time a great majority of studies dealt with the mechanism of frost damage to concrete after hardening. The freezing of concrete at an early stage before it lost plasticity is less well understood whereas the conditions of concrete hardening at this stage are of vital importance to the formation of its structure and to physico-mechanical properties.

Without dwelling on the existing theories of frost damage to hardened concrete we go over to an analysis of causes for structural deterioration when concrete freezes at an early age. Studies that we conducted at NIIZhB (Research Institute of Concrete and Reinforced Concrete) in Russia found a number of regularities in the destructive phenomena that arise in concrete when it freezes, thaws and then hardens at an above-zero temperature.

- Causes of deterioration of concrete may be divided into the following three groups:
- Conversion of water into ice that causes an increase in volume and great internal pressures.
- Mass transfer as the concrete freezes, which leads to the redistribution of moisture and the formation of large ice segregations.
- Disintegration of the whole multicomponent system due to the mass transfer and sedimentation phenomena, which weakens the bond between the coarse aggregate particles and the paste constituents of the concrete.

Deterioration of Concrete in Freezing. The conversion of water into ice whose volume increases by 9% causes high internal pressures as described above. When frozen immediately after mixing, concrete has no elasticity and cannot resist the frozen moisture that fills its intergrain space. This pressure forces solid particles apart. It can be clearly seen in the variation of strains in concrete frozen immediately after placing (Figure 2.3). A specimen contracts as it is cooled down to 0°C. There is an abrupt change in the strains within the temperature range between 0 and –1°C, which characterizes expansion of the concrete due to the conversion of mechanically bound moisture into the solid phase. The strains gradually decrease as the concrete is cooled further.

So in fresh concrete, at temperatures from 0 to –2°C, most of the moisture (up to 90%) turns into ice* and causes a high internal pressure in the material, which cannot be resisted by the plastic system without deformation and an increase in volume.

When the temperature of the frozen concrete rises, its volume increases again due to expansion of all its constituents. As the temperature crosses the zero point and ice starts to melt, the internal pressure drops as a result of the contraction of the liquid volume and of the opening of channels for migration of the gaseous fraction that was blocked in voids by ice. The loose system starts to harden quickly when the temperature is above zero. The strength of such concrete remains lower than that of unfrozen concrete that hardened at normal temperature and humidity.

The expansion of water in freezing also damages concretes that had attained some strength by the time they froze. However, as concrete hardens and its elastic characteristics improve, the damage done to its structure by freezing diminish. The

---

* The process of ice formation is somewhat different for the freezing of concrete with a W/C ratio lower than that shown in Fig. 2.2.

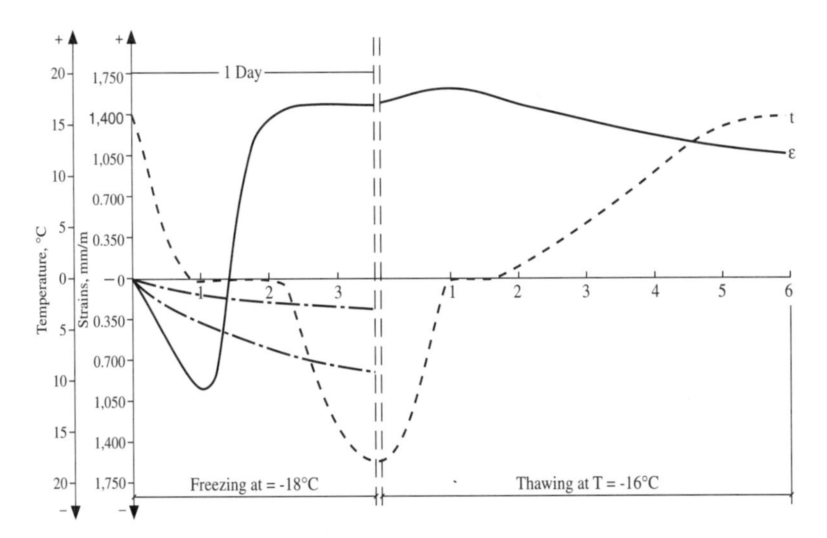

FIGURE 2.3 Strains in concrete frozen immediately after mixing and in the process of thawing. ---- Strains in freezing and thawing in a humid environment, -.- Strains in hardening under normal conditions, -…- Strains in hardening in dry air

reason is a reduction in the amount of mechanically bound water and this noticeably affects strains. Thus, concrete specimens frozen 14 and 17 hours after hardening under normal conditions practically had no expansion strains when the temperature rose above 0°C. Changing porosity of the material also helps. Macropores that get rid of mechanically bound water as it is bound chemically and physically or evaporates are filled with air and become reservoirs where the air compresses when the liquid phase freezes and so the capillary walls are not ruptured to any great extent. Air-entraining admixtures are very useful for reducing the deterioration of concrete. They increase the durability of the material and of concrete structures subjected to cycles of freezing and thawing.

Structural Deterioration Due To Mass Transfer. Mass transfer causes a great structural deterioration of concrete of physical nature. When concrete is mixed in a mixer, water spreads evenly over the whole mass. Moisture may redistribute during transportation, handling, and storage of mixed concrete. So some parts of the mixture may become excessively saturated with water and large structural faults can be expected there in freezing. But even if the moisture was distributed uniformly over the concrete mass after placing and compaction, it starts to move after freezing and the appearance of a temperature gradient in the structural member.

The member starts to cool from the surface of concrete, with the cold front gradually spreading into the inner core. Since partial pressure is lower in the area of lower temperatures, the liquid phase starts to migrate to the colder area. Having reached the zero isotherm, it freezes and expands forcing the grains of cement and aggregate apart in fresh concrete or rupturing the crystal skeleton in the concrete that began to harden. Ice inclusions are formed in the pores and cavities.

Water diffuses to the ice inclusions from thinner capillaries (including those inside gel) because these inclusions are cold sinks inside the material. Another reason is as follows.

The reaction of cement with water is exothermal and involves heat evolution. Although frozen concrete contains little water and the latter has a low reactivity, still the reaction does take place and some exothermal heat evolves raising the temperature on the surfaces of cement grains. This causes the moisture to migrate to the colder area, i.e., to ice inclusions, where it changes into mechanically bound water from adsorptionally bound one and freezes. The ice inclusions increase in volume and gradually form ice segregations which are often visible, particularly in the layers close to the concrete surface.

Several factors affect the diffusion of moisture when concrete freezes, including the temperature and the freezing rate, the porosity, and the water–cement ratio of the concrete. Without dwelling on the last two factors whose role was described in literature in sufficient detail we will discuss briefly the temperature and the freezing rate of concrete.

The temperature of freezing is particularly important to fresh concrete. While about 92% of mixing water turns into ice at –5°C, this value rises to 96% at –45°C (Figure 2.2). Consequently, very little adsorptionally bound water remains liquid in the concrete. The amount of the moisture that can diffuse to ice inclusions is so low that they do not grow at all at low subzero temperatures. We have not found any important changes in ice segregations even after curing freshly frozen concrete in the frost at –20°C for three months.

This phenomenon was not observed either in concretes that had attained some strength by the time of freezing since their microcapillaries lost continuity of moisture filling and the diffusion of the liquid became practically impossible. True, diffusion of water vapor takes place there but it is not important and does not affect the growth of ice segregations to any great extent.

The picture is different at subzero temperatures close to 0°C. Concrete has enough mechanically bound water and weakly adsorptionally bound water (e.g., only 35% of water turns into ice in concrete with W/C = 0.5 frozen at –2°C when its strength was 15% R_{28}) capable of migrating. Placed concrete cools down much slower at a temperature close to 0°C and that also helps since it creates favorable conditions for movement of moisture before it freezes. When placed concrete cools down and freezes slowly, the moisture is redistributed to a greater extent. It migrates to the cold front and to the ice inclusions. This promotes segregation of ice and reduces the density of the concrete.

The migration of water in hardening and hardened cement mortars or concretes may cause failure of a structure when the temperature gradient is high.

An example is splitting of glass blocks in a window opening during construction of a marketplace at a city in Siberia. Ice growths appeared from the outside of the building in mortar joints between glass blocks. The growths built up rapidly in the frost that reached –45°C. Inspection of the facility and an analysis of the splitting have shown that the reason was migration of moisture. Humidity in the room was high when it was being finished. The temperature drop between the air inside and outside was 60 to 65°C. The mortar in the joints 10 to 15 mm wide and 150 mm deep (the thickness of a glass block) was placed in summer and must have been highly porous. In winter, the temperature of the part of each joint adjacent to the interior surface of the window was above zero and of that adjacent to the exterior surface, below zero. The zero isotherm passed at about one-third of the distance from the exterior surface and the mortar in the joints acted as a kind of a pump that delivered moisture from the warm area to the cold one. There the moisture froze and expanded causing high internal stresses in the mortar, which cut off the exterior halves of the glass blocks. Internal glazing was recommended to prevent access of the moisture in the room to the mortar joints and to stop this destructive process.

Even a brief analysis of the moisture migration in concrete during freezing shows that this factor is of key importance and may cause an irreversible structural damage. When ice segregations melt, they leave behind cavities and directed capillaries thus degrading the quality and durability of concrete structures.

Disruption of the solidity of concrete. The solidity of a conglomerate system, such as concrete, is the basis for its proper behavior and any disruption may cause disintegration of the material.

When fresh concrete freezes, the interaction between coarse aggregate particles and paste is impaired by the mass transfer. The particles of coarse aggregate in normal weight concrete are usually denser and conduct heat better than the cement paste. So their temperature drops down faster when the concrete cools and they attract moisture from the core. As it freezes at the interface with the aggregate, the water forms an ice film that continues to grow due to diffusion of the moisture from capillaries. The film may be 1 mm thick or more.

The ice film at the interface of the paste and of the coarse aggregate particles benefits sedimentation, too. The ice film originally formed by sedimentation may also grow under favorable conditions due to the influx of moisture from capillaries. As it grows in volume, the ice film widens the gap between the aggregate and the paste not only breaking their contact but also loosening the structure of the concrete in general. The film disappears when it melts and the resultant air gap disrupts adhesion of the components. Even if the concrete hardened under favorable conditions in terms of temperature and humidity, the adhesion remains very low after thawing and the aggregate particles can be easily separated from the paste in the concrete. Traces of the frozen water can be seen clearly on the interface surface of the cavity in the form of frostings. It is characteristic that, as the density of the aggregate lowers, the traces of moisture frozen at the interface become less evident (Figure 2.4).

The situation is different when frozen concrete contains porous aggregates. The aggregate particles absorb moisture and no water film is formed at their interface with the cement paste as a result. So the concrete deteriorates less. The solidity of concrete with porous aggregates, however, depends largely on the concrete's type and structure.

When aggregate particles contain a lot of microcapillaries (porous limestone is an example), they are saturated with moisture that does not freeze at 0°C because of capillary forces. As the concrete cools down, the moisture starts to migrate to the cold front and an ice film is formed on the particle side that faces the zero isotherm. The film grows rapidly due to suction of moisture from the interior layers of the

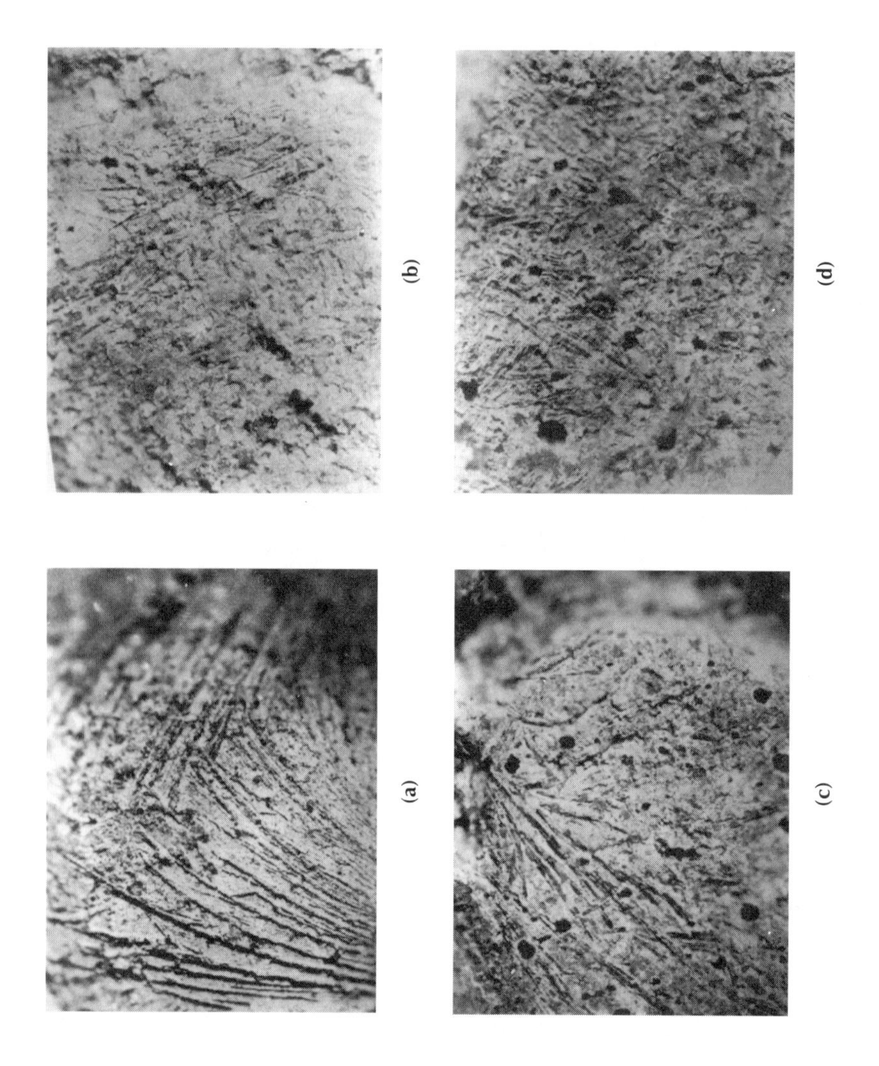
(a)
(b)
(c)
(d)

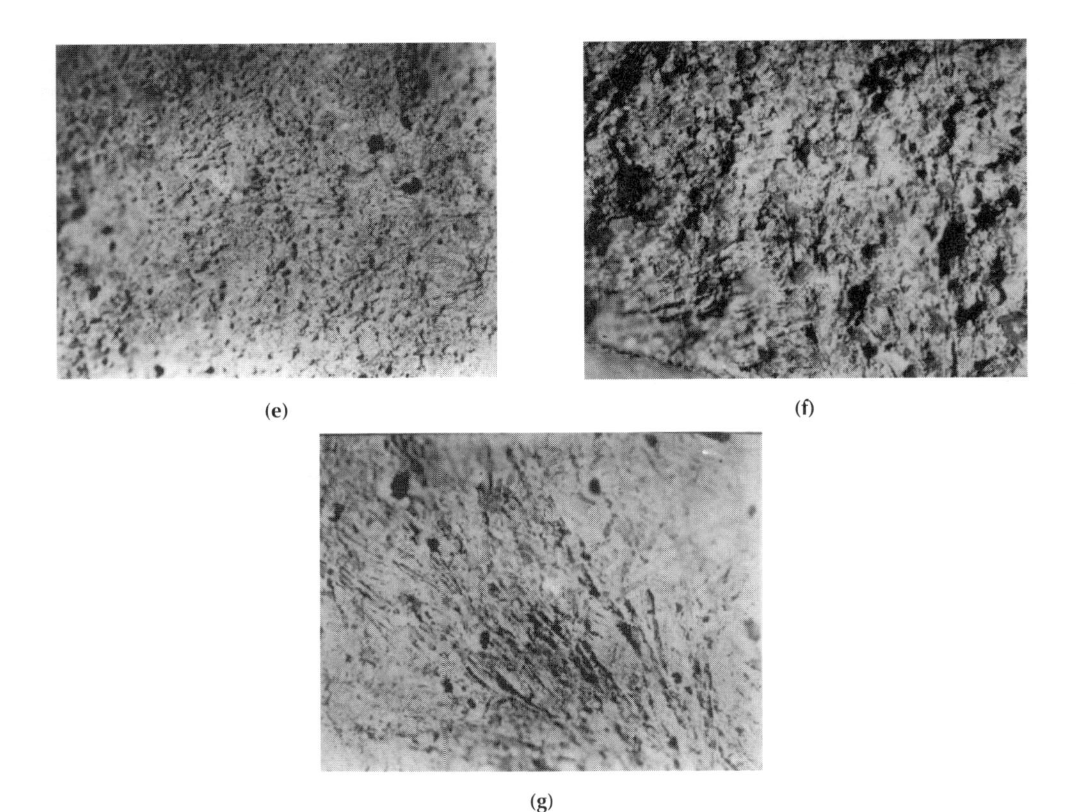

FIGURE 2.4 Freezing of water in concrete at the interface with coarse aggregate particles of different density and moisture content. (a) metal ball (γ_o = 7.8 g/cm^3, moisture = 0%), (b) granite (γ_o = 2.68 g/cm^3, moisture = 0.4%), (c) pebbles (γ_o = 2.64 g/cm^3, moisture = 1.2%), (d) limestone (γ_o = 2.50 g/cm^3, moisture = 2.5%), (e) red brick (γ_o = 1.70 g/cm^3, moisture = 18.7%), (f) slag pumice (γ_o = 1.53 g/cm^3, moisture = 15.5%), (g) expanded clay (γ_o = 0.5 g/cm^3, moisture = 16.5%).

particle and from the adjacent cement paste. The thickness of the film may reach several millimeters and it can disrupt the structure of the concrete considerably. These particles are particularly dangerous in the layers close to the surface of concrete, where bulges appear on the surface of the structure made of such concrete due to growth of large ice segregations. The bulges easily break in thawing.

Porous aggregates, such as expanded clay, pumice, or volcanic slag, can reduce considerably the frost damage to concrete at an early age. The absorption of a part of moisture from cement paste by particles with a multitude of microcapillaries and pores before freezing reduces the water–cement ratio. The amount of mechanically bound water in the paste falls sharply. When the concrete freezes, the water remaining in the intergrain space does not disrupt the structure too much because the expansion of the moisture is compensated by a great number of air-filled pores.

The absorbed moisture also freezes in the aggregate's macropores but it does not have a destructive effect because there are many pores that are not filled with water. After it thaws, the concrete with porous aggregates continues to harden rapidly and develops a higher strength than normal weight concrete under the same conditions. At this stage, as the cement paste dehydrates and a moisture gradient appears, moisture migrates from aggregate particles creating better conditions for hardening.

It should be emphasized, however, that in concretes with a high W/C ratio the porous aggregate increases the frost damage rather than prevents it. Since the aggregate is highly saturated with moisture, its particles break when they freeze and the quality of the concrete deteriorates dramatically.

2.4 PRECURING OF CONCRETE BEFORE FREEZING

Studies conducted by many researchers (S. A. Mironov, N. A. Moshchansky, A. A. Shishkin, I. Blondel, M. Durier, O. Graf, H. Scofield, A. Velmi, G. Meller, I. Lis, A. Nykenen, and H. Tatill) have shown that concrete should develop a certain strength by the time it freezes to avoid frost damage to its structure and properties. It should be emphasized once again that this strength development by the time of freezing makes the concrete a capillary-porous material containing little mechanically bound water that adversely affects concrete in freezing.

The time during which concrete must be cured before it is allowed to freeze may be different and varies from 12 hours to 28 days according to the type and class (grade) of concrete, type of cement, water–cement ratio, the environment where a concrete structure will be maintained, and some other factors. It corresponds to a strength of 20 to 100% R_{28}. The strength that allows concrete to freeze without degrading its quality is called "critical strength."

All national and international guidelines, standards, and specifications prohibit to freeze concrete before it attains its critical strength, which ranges from 25 or 30% to 50 or 80% R_{28} in different countries. A correct choice of the critical strength is of great technical and economic importance.

The Research Institute of Concrete and Reinforced Concrete of Russia (NIIZhB) conducted extensive studies under my guidance to determine effects of frost on concrete, including determination of the critical strength. Here are results of some experiments with freezing concretes of different strengths ranging from 0 to 100%

R_{28} at temperatures of –5, –20, and –50°C. Cubes of 100 × 100 × 100 mm cast of Portland cement concrete with a slump of one to three cm were tested within four hours after thawing at 20°C.

Concretes frozen immediately after mixing had practically no strength gain, except those that were held at –5°C. Their strength was 8% R_{28} after thawing. Similarly, concretes that were cured for 28 and 90 days at –10, –20, and –50°C had strengths within 1.5 to 4.5 kg/cm2 after four-hour thawing, which made 0.46 to 1.4% R_{28}. These results agree quite well with the data on the phase state of water in frozen concrete set forth above. Indeed, the adsorptionally bound water that remains liquid has a low reactivity and so its reaction with cement is very slow and the concrete practically does not develop any strength in the frost.

The studies proved that there was no hardening in the frost if concrete did not contain enough water in the liquid phase. Many experiments that I made demonstrated that cement did not react at all with water in the solid state (ice).

The picture was somewhat different when we froze concrete that had attained some strength by that time. Literature contains very contradictory information but a number of authors agree that concrete still develops some strength in the frost.

Our studies found that concretes frozen when their strength was 15 to 45% R_{28} and tested after four hours of thawing had the greatest strength gain. The strength of these concretes increased by 23 to 28% after the specimens had been kept in a cooling chamber for 28 days (Figure 2.5). Concrete frozen when its strength was 70 to 80% or more practically did not increase its strength after 28 days of curing at –5°C. The reason was the presence of newly-formed dense films around cement grains that had developed a strength of 70% or more by the time of freezing. This considerably impaired the influx of moisture to the unhydrated parts of the grains and even more so because the viscosity of the moisture was higher (the viscosity of water was 1.789 P as compared with 1.005 P at 20°C and 0.550 P at 50°C) due to aggregation of molecules and a slower motion. According to T. Powers, the viscosity of water passing through cement gel with pore diameters in the order of 2×10^{-7} cm is about 50,000 times higher than that of water at normal temperature.

Concretes frozen with a lower strength did not undergo such structural changes. They had more large pores, the diffusion of moisture was faster, and the films of new formations around cement grains were thinner and less dense. This favored to some extent faster hardening processes and an increase in strength in the frost.

Curing of concretes in cooling chambers at –20°C (Figure 2.6) and –50°C (Figure 2.7) showed that their absolute and relative strength gains at different ages decreased when temperature dropped. Long curing of concrete specimens at low subzero temperatures down to –50°C did not change their strength as compared with short freezing. Curiously, when the same concrete was cured at –20°C for seven and 90 days, its strength decreased in some experiments by 3 to 15%. The reason must have been more favorable conditions for ice segregation at a microscopic scale at –20°C than at –50°C, since the mass transfer goes on faster at this temperature due to a large amount of moisture in the liquid phase.

Experimental studies of cement hydration and its relationship with the strength of concrete were quite interesting. The results proved that a higher degree of hydration did not mean that such a concrete had a high strength. When hydration was

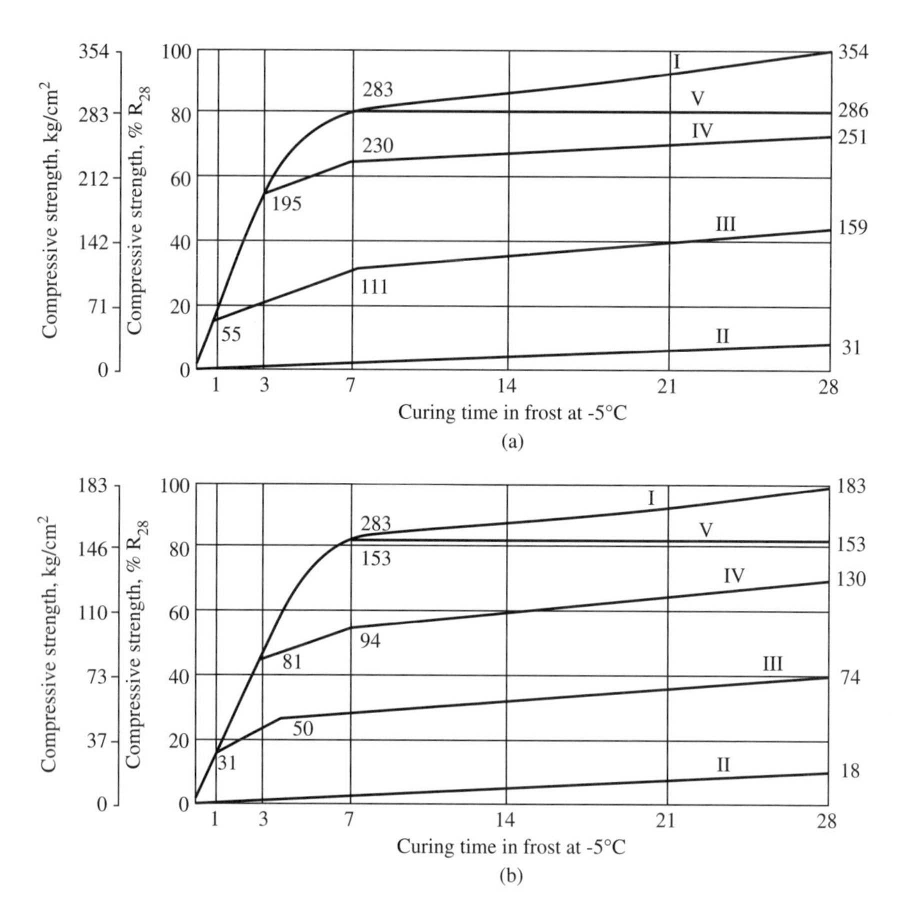

FIGURE 2.5 Strength development of concrete cured in the frost at T = –5°C (when tested after thawing for four hours). A — Portland-cement concrete with W/C = 0.46, B — Portland-cement concrete with W/C = 0.72, I — hardening under normal conditions, II — freezing immediately after mixing, III — freezing at an age of one day, IV — freezing at an age of three days, V — freezing at an age of seven days

compared with strength development of hardened cement paste after thawing, it was found that specimens frozen immediately after mixing had hydrated to practically the same degree as specimens frozen with a strength of 70% R_{28} but the compressive strength of the former after thawing was the lowest. Consequently, the processes that go on in the frost are mainly dissolution of minerals of the cement clinker, saturation of the liquid phase with hydration products, and formation of a colloidal mass. Along with a better dissolution of lime at low temperatures, these are the reasons for a higher degree of cement hydration in concrete at an early age. Gel films around unhydrated parts of cement grains remain penetrable for water for a long time. The water diffuses through gel pores and causes new portions of the cement to hydrate. There appears to be no proper conditions at subzero temperatures for the formation of crystal intergrowths whose strength determines the strength of hardened cement paste.

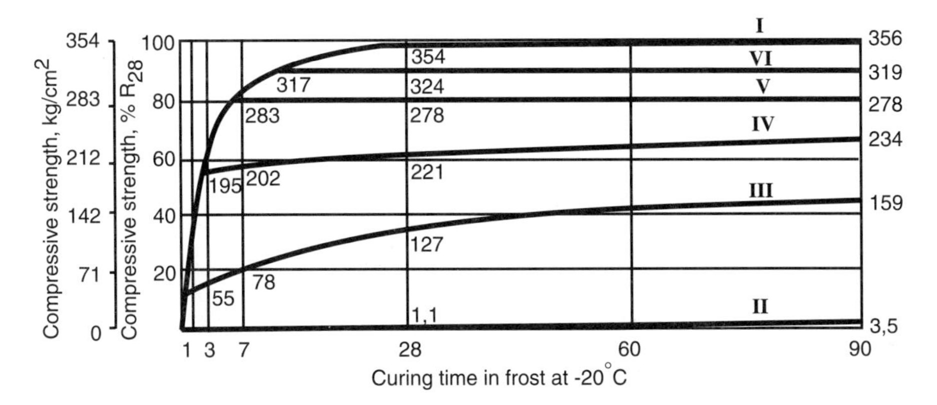

FIGURE 2.6 Strength development of concrete with W/C = 0.46 cured in the frost at T = –20°C. I — hardening under normal conditions, II — freezing immediately after mixing, III — freezing at an age of one day, IV — freezing at an age of three days, V — freezing at an age of seven days, VI — freezing at an age of 14 days.

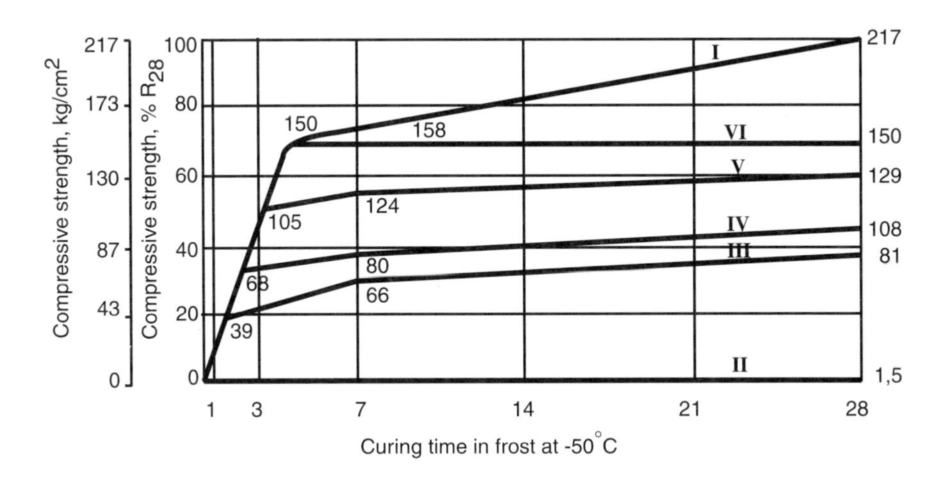

FIGURE 2.7 Strength development of concrete cured in the frost at T = –50°C. I — hardening under normal conditions, II — freezing immediately after mixing, III — freezing at an age of one day, IV — freezing at an age of three days, V — freezing at an age of seven days, VI — freezing at an age of 14 days.

Special investigations identified the role played by cooling and freezing of concrete, and by conditions wherein it thawed, in kinetics of the strength development. Such a concrete always passes three stages that include cooling to 0°C, freezing, and thawing. The importance of each stage to the strength development is undoubtedly of interest.

In our experiments, concrete specimens of 100 × 100 × 100 mm were cooled at different rates to –1°C whereupon they were thawed in a chamber of normal conditions to 2°C and tested as soon as that temperature was reached and four or five hours later, i.e., after the concrete attained the ambient temperature (20°C). Some specimens of the same concrete were cooled to –10 and –20°C at different rates. After their strength was determined, samples were selected for a chemical analysis and dehydrated with alcohol at once. Thus the degree of hydration was found according to the amount of $Ca(OH)_2$ evolved.

An analysis of the data (Tables 2.1 and 2.2) thus obtained revealed that each stage had some significance and its importance in the concrete strength development may increase or decrease depending on conditions. When concrete with a strength of 8 to 30% was cooled rapidly, the strength had no time to increase at this stage and remained about the same when the specimens were thawed to 2°C (the average strength gain was 2% R_{28} during the time of cooling and thawing). Typically, keeping rapidly cooled concrete for three days at –20°C did not help to increase its strength as compared with specimens cooled to –1°C in one hour and thawed to 2°C in one hour and, on the contrary, resulted in a lower strength gain or in a decrease in strength. This can be easily explained in terms of the phase state of mixing water in concrete at this temperature and of the above-described destructive phenomena.

Data on strength development of concrete thawed to normal temperature for four or five hours are of most interest. This stage clearly stood out and was characterized by a rapid development of the strength of the material in all cases regardless of the freezing temperature.

Concretes that had attained a sufficiently high strength at 2°C by the time of freezing did not develop it further and, moreover, lost 1 to 6% of it in all cases. However, after four or five hours of thawing the strength also increased rapidly reaching the initial value (before freezing) or somewhat exceeding it.

The strength development data agreed quite well with the chemical analysis of $Ca(OH)_2$ content in percent of that in dried specimens (Table 2.2).

When analyzing phenomena that take place in concrete during freezing and thawing and related hardening processes, we must dwell on a special significance of natural conditions of concrete curing as compared with conditions of cooling chambers. The basic difference is that, in the natural environment, the ambient temperature is very unstable and its variation may be as high as 20°C even within one day. Temperature conditions of cooling chambers, on the contrary, are quite stable, which is very important for clarifying the influence of individual factors on the processes taking place in concrete.

Thus, natural conditions are unacceptable for experiments to determine the effects of the environment on characteristics of concrete since they will always have an additional variable, i.e., temperature. The variability of the temperature factor entails changes in kinetics of the phase state of water, of mass transfer, and of chemical reactions. This may be the reason why some researchers saw concrete develop strength in the frost and drew wrong general conclusions when they experimented in the natural environment.

The foregoing was proved by studies conducted by NIIZhB. Concrete specimens were divided into two batches. The first was placed outside after casting and stayed there for three months (from Jan. 3 to March 30, 1996, in Moscow). The second batch was kept in a cooling chamber at –20°C during the same period. The air

TABLE 2.1
The impact of freezing temperature and thawing time on the strength of concrete

Curing before freezing in hrs.	Concrete strength before freezing in kg/cm^2 and % R_{28}	Freezing to –1°C		Freezing to –10°C		Freezing to –20°C		Freezing at –20°C for 3 days	
		Thawing in chamber of normal storage							
		to 2°C	for 4 to 5 hrs	to 2°C	for 4 to 5 hrs	to 2°C	for 4 to 5 hrs	to 2°C	for 4 to 5 hrs
12	28/8.1	37/10.7	81/23	36/10.4	60/17	42/12	58/17	29/8.4	53/15
24	106/31	113/33	125/36	99/29	127/36.5	100/29	126/36	104/30	131/38
48	190/55	177/51.5	200/58	184/53	226/65	185/53	207/60	181/52	199/58
72	236/68	221/65	239/69	228/66	257/74	218/63	236/68	211/61	237/68

Note: Concrete with M-600 cement, R28 = 346 kg/cm2

TABLE 2.2
The impact of freezing temperature and thawing time on the degree of hydration of cement

Curing before freezing in hrs.	Quantity of $Ca(OH)_2$ in % of dried specimen			
	before freezing	after freezing to –1°C and thawing to 2°C	after freezing to –20°C and thawing during	
			1 hr to 2°C	4 or 5 hrs to 16–18°C
12	0.74	1.19	1.20	1.32
24	2.62	2.89	3.10	3.26
48	3.84	3.97	4.15	4.28
72	5.66	5.67	5.90	6.08

temperature outside varied during that time from –32 to 6°C. Average daily above-zero temperatures of 2 or 3°C were observed during one day late in February and after March 22.

The results (Table 2.3) clearly demonstrate that natural conditions are much more favorable for concrete hardening than rigid temperature conditions of cooling chambers.

The strength of concretes frozen and cured in the cooling chamber was much lower than that of the concretes which were kept in the natural environment. In the latter, temperature variation caused continuous changes in the phase state of water and accumulation of hydration products. Better dissolution of lime at low above-zero temperatures made the hydration of cement grains go deeper.

The findings of the experiments lead to the following conclusion: essentially concrete does not develop any strength at a temperature lower than –5°C. At subzero temperatures, minerals of cement clinker dissolve, the liquid phase is saturated, and hydration products accumulate in concrete at different rates depending on a particular temperature value. Strength increases as the concrete thaws at an above-zero temperature due to faster crystallization and formation of the skeleton of hardened cement paste.

So concrete develops its strength before freezing and after thawing, and does not increase it for all practical purposes in the frost. Assertions of some authors, e.g., I. A. Kireenko, that the strength of concrete can grow fast in the frost are wrong and must have been the result of errors in the methodology of investigations.

2.5 EFFECTS OF EARLY FREEZING ON CONCRETE PROPERTIES

The properties and durability of concrete frozen at an early age may greatly deteriorate due to structural defects. Underdeveloped strength, low water impermeability and frost resistance, and low bond with the reinforcement are practically irreparable. Thus, compressive strength of concretes with W/C = 0.6 or more may be lower than a normal level by as much as 40%. Frost resistance of concrete designed for 200 cycles of freezing and thawing in water saturated condition may drop to 10 of this value or lower. Impermeability to water designed to withstand 8 atm of pressure may decrease to 1 atm.

Early freezing of concrete is particularly harmful to its bond with the reinforcement in reinforced concrete structures. Thus, when a structural member is reinforced with smooth bars, the bond may weaken by 80%. With deformed bars, the bond may decrease by 20%. Due to mechanical adhesion of concrete to projections on the bars the bond does not drop too much. But because of the loss of contact with the steel surface a gap produced by melted water film remains between the concrete and the rebars. Water vapor, air oxygen, and corrosive agents may diffuse into the gap from the environment and to start corrosion of steel. Naturally, the durability of the reinforced concrete member may decrease considerably.

All this implies that the freezing of concrete that did not reach structural maturity adversely affects its properties and should not be allowed to happen.

TABLE 2.3
Variation of concrete strength in the frost in a cooling chamber at –20°C and in the natural environment of winter in 1965–1966

Curing time before freezing in hrs.	Concrete with W/C = 0.46			Concrete with W/C = 0.72		
		in thawed state after 90 days of curing			in thawed state after 90 days of curing	
	before freezing	in cooling chamber at –20°C	in natural winter conditions	before freezing	in cooling chamber at –20°C	in natural winter conditions
0	0	1.3/0.5	214/62	0	0	124/52
0.5	12/3.5	24/7.0	217/63	13.5	20/8.5	120/50
1	96/28	139/41	307/89	53/22	74/31	176/74
2	188/54	207/60	308/89	92/38	115/48	176/74
3	205/59	225/65	313/91	121/51	138/58	206/87
7	255/74	248/72	312/90	175/74	176/74	229/96
28	345/100	332/97	337/98	238/100	235/99	243/102

There are cases in building practice when concrete structures are maintained in frozen condition. Frozen concrete is quite strong due to adhesive action of mechanically bound water turned into ice. But it takes place to a certain limit of the water content in the concrete, whereafter its strength is determined by that of ice. When it thaws, the load-carrying capacity of the concrete and the durability of the structure drop dramatically. As was mentioned above, regulations of the Russian Federation established a critical strength of concrete, which should be reached before the concrete may be allowed to freeze without serious adverse effects on its structure and properties.

Concrete of any class up to 12.5 may be frozen when it attained a compressive strength of 50%. Concrete of a class up to B22.5 may be frozen when its strength is 40%. Concrete of any class over B22.5 may be frozen when it developed a strength of 30%. The structural members that must meet high specifications for frost resistance should have the strength of their concrete equal at least to 80% of the designed value by the time they can be frozen.

Strength characteristics established by regulations are indirect characteristics. The above-listed values do not mean that, once they are reached, the structure of concrete cannot be broken by frozen water. When concrete attains these strength values, it contains much less mechanically bound water that turns into ice at 0°C. Spent on physico-chemical processes and evaporation, it leaves behind pores and capillaries filled with air. Due to good compressibility of the latter, they withstand pressure of the remaining water caused by freezing and thus protect the structural skeleton made up of new formations produced by cement hydration. Air bubbles introduced into the concrete by adding air-entraining agents to the concrete mixture also have a beneficial effect on this process.

2.6 CONCRETE IN THE FROZEN STATE AND ITS BASIC PROPERTIES

Concrete and structures thereof are maintained at a subzero temperature most of the year in the Arctic and permafrost regions. There have been cases when placed concrete was frozen at an early age and never thawed again. The durability of a number of structural members and of whole structures operating under these conditions depends on their properties in the frozen state, such as strength and deformability, including creep.

That problem was studied by researchers in many countries, including B. G. Skramtaev, S. A. Mironov, Schultz and Altner, Monfore and Lentz, and Moniel. According to their data, frozen concrete has a high strength that may reach sometimes 200% of its design value. The bending-tensile strength of prisms measuring 100 × 100 × 100 mm increased by 190% when they were frozen to –150°C. Concretes in the frozen state have a higher dynamic modulus of elasticity.

According to N. V. Sviridov, cube strength, prism strength, and the static modulus of elasticity increase when temperature drops to –50°C. Cube strength may rise by as much as 152%, prism strength by 209%, and the static modulus of elasticity by 126%. The moisture content of the concrete is of great importance there.

Our studies conducted by O. S. Ivanova at NIIZhB have shown that the water–cement ratio and initial water content also play a part. Concretes with W/C = 0.72 frozen at an early stage had a lower strength in the frozen state than those with W/C = 0.44. As temperature drops, strength increases considerably, particularly at temperatures below –20°C due to a change in the ice content of the concrete.

So the properties of frozen concrete change to a great extent and need an extensive research, particularly deformation properties that may be affected by ice. The increase in strength and improvements in other physico-mechanical characteristics of concretes in the frozen state are the result of the adhesive action of ice. This increases the strength of the concrete's skeleton. So the cementing properties of ice can be used to advantage in plain and reinforced concrete structures erected under the Arctic and permafrost conditions.

The NIIZhB studies used concretes with a strength of 300 kg/cm^2 mixed with Portland cement, the slump of the mixtures ranging from 4 to 6 cm. After they were cast, cubes of 100 × 100 × 100 mm and prisms of 100 × 100 × 400 mm were covered with a plastic film and were frozen for seven days in a cooling chamber at temperatures from –26 to –30°C immediately after casting, after one day, and after two days of hardening in the natural environment at 12 to 14°C and a relative air humidity of 55 to 60%. In addition, one series of specimens was frozen at the age of 28 days after curing in normal conditions.

To identify the role played by cement, specimens of the same size were made. They modeled concrete but contained crushed quartz sand instead of cement in the same amount and with the same dispersity, its specific surface being 3000 cm^2/g.

After they were cast, the specimen models were covered by the plastic film and frozen in the cooling chamber. Specimens were also made of pure ice whose water was the same that was used for concrete mixing. To reduce internal stresses and produce whole ice specimens, the water was poured in molds and frozen in layers.

The specimens were tested after seven days of freezing. To minimize the temperature rise in the frozen specimens, the latter were well insulated on all sides with layers of foamed plastic 3 cm thick. Cold metal sheets 40 mm thick were installed between the plates of a press and supporting faces of specimens when cubes were tested for compression. Prisms made of ice were tested in a press placed in the cooling chamber at –28°C. A special technique was developed for determining strain characteristics to minimize the effects of temperature and other factors.

An analysis of the studies demonstrated that the cube strength of concrete in the frozen state depends on the time of precuring before freezing at an above-zero temperature. The strength of concrete frozen immediately after casting and kept at –28 to –30°C for seven days was 83% of its design value. The reason for this high strength was the cementing action of ice because more than 94% of mixing water turned into ice at –30°C. Of course, in addition to the strength of the ice itself, the strength of the concrete also depended to a great extent on the nature of the ice freezing together with solid particles and on the roughness of the surface and on the density of coarse and fine aggregate particles.

Curiously, when we tested concrete specimens frozen immediately after casting, we did not see the usual brittle failure and shattering of compressed concrete. The cubes failed as if they were made of a highly plastic material; they preserved their shape and were crumpled — the evidence of high viscosity.

The strength of concrete frozen after curing for one day at an above-zero temperature was 116% of its design strength in the frozen state. The strength acquired by concrete while it was cured in the above-zero temperature conditions favored its further development in the frozen state. Its strength was determined by that of the structural skeleton that began to form and by the cementing action of water when it turned into ice. This is the reason why the strength of the concrete in the frozen state was so high. Tests of the cubes produced some cracking showing that the material was rather brittle.

The strength of concrete frozen after curing for two days at an above-zero temperature reached 128% of its design value in the frozen state. Testing produced brittle failure with splitting and cracking.

The strength of concrete frozen after 28 days of curing in normal conditions was 146% of the design strength in the frozen state. The failure of the specimens was brittle with shattering.

Tests of specimen models in the frozen state showed a strength of 306 kg/cm^2. It was almost the same strength that was found in the concrete specimens frozen immediately after casting (291 kg/cm^2). The nature of failure of the cubes proved that they were highly plastic: they were crumpled rather than shattered. The test results and the strength of the specimen models, which was close to that of the concrete specimens frozen immediately after casting, enable us to say that the main cementing material in both was water turned into ice and its content in the two materials was the same. So the strength of fresh concrete in the frozen state depends on that of ice.

The strength of test specimens made of pure ice varied over a broad range from 11 to 86 kg/cm^2. Similar results were obtained by other researchers who tested ice specimens.

A special technique had to be developed for experiments with ice specimens. If you try to freeze water that filled a mold to the brim, you will not get whole ice specimens. When the water freezes, the specimen cracks and falls apart as it is taken out of the mold. The reason is high stresses that arise when the extreme water layers freeze and make a shell around liquid water in the core. The temperature of the water being 0°C and that of the air –28°C, the water surface gets covered with a multitude of small crystals that grow rapidly towards each other. The ice thickness increases at the rate of 5 mm/hr. The surface of the specimen becomes covered completely with ice that closes the whole volume of the still unfrozen water. Expansion of water in a closed volume produces internal pressure that may be as high as 2000 or 2500 kg/cm^2. This cracks the outer layers of the ice shell.

To relieve internal stresses, to prevent disintegration of ice specimens, and to reduce the spread of strength values, water should be frozen in molds layer by layer for any type of specimens. The thinner the water layers the lower stresses will arise in a specimen. It can be clearly seen from strength properties of our ice specimens. Thus, 100 × 100 × 100 mm cubes made in layers of 33 mm were whole but small bulges with some shallow cracks could be seen on their exposed surfaces. Their compressive strength was 52 kg/cm^2. The same specimens made in layers 10 mm thick had not surface cracks and their strength reached 89 kg/cm^2.

The studies allowed to get a clearer idea of the processes in concrete frozen at an early age and of their effects on its strength.

In addition to strength, deformability is also very important to such concretes. Prism specimens were made of the same concrete as the ones for strength tests. They were frozen immediately after casting, after one day, and after two days of precuring at 12 to 14°C. Some specimens were frozen after they hardened for 28 days in normal conditions.

The tests demonstrated that longitudinal compression strains of all concrete specimens in the frozen state after precuring were higher than those of unfrozen specimens under the same loads. They increased as the precuring time before freezing got shorter (Figure 2.8).

Deformation of concrete specimens frozen immediately after placing rose considerably even for stresses of $0.45R_p$. In other words, total deformations increase due to an increase of the plastic component. It is evidence of a high plastic flow of ice under loading because it was the main binding material in the test specimens. So, for a stress of $0.4R_p$ total deformations of freshly frozen concrete were 50×10^{-5} and they grew to 140×10^{-5} when the stress was $0.46R_p$.

Strains of concrete specimens frozen after one day of curing were somewhat different in values and nature. Their total deformations were 130×10^{-5} for a level of stresses of 0.6Rp.

Longer time of curing at an above-zero temperature before freezing affects substantially values of plastic deformations of frozen specimens. Total deformations of specimens frozen after 2 and 28 days of precuring dropped and were 68×10^{-5} and 67×10^{-5}, respectively, for the $0.6R_p$ stresses. They were 63×10^{-5} under the same stresses in unfrozen check specimens. The growth of total deformations due to higher plastic ones under loads of $0.6R_p$ showed the role of the amount of ice in frozen concrete specimens and the destructive processes that are triggered in concretes by freezing.

Deformations of frozen specimen models and ice proved their high plasticity under load. When the level of stresses was $0.6R_p$, total deformations of the specimen models reached 230×10^{-5} and those of ice under stresses of $0.4R_p$ were 260×10^{-5}.

The prism strength of frozen concrete studied on 100 × 100 × 400 mm prisms increased with the time of curing at an above-zero temperature before freezing (Table 2.4). The strength of the specimens frozen at –26 to -30°C for seven days immediately after casting was 244 kg/cm^2 or 54% R_{28} of concrete frozen after 28 days of hardening under normal conditions or 89% of the prism strength of concrete that had hardened for 28 days under normal conditions. The nature of failure of concrete prisms frozen immediately after casting revealed high plasticity of the material. The specimens preserved their shape after testing because they were only crumpled.

The same plastic properties were characteristic of the specimen models. Their prism strength was 280 kg/cm^2.

The prism strength of concrete specimens frozen after one, two, and 28 days of precuring was 365, 405, and 450 kg/cm^2, respectively, while their strength before

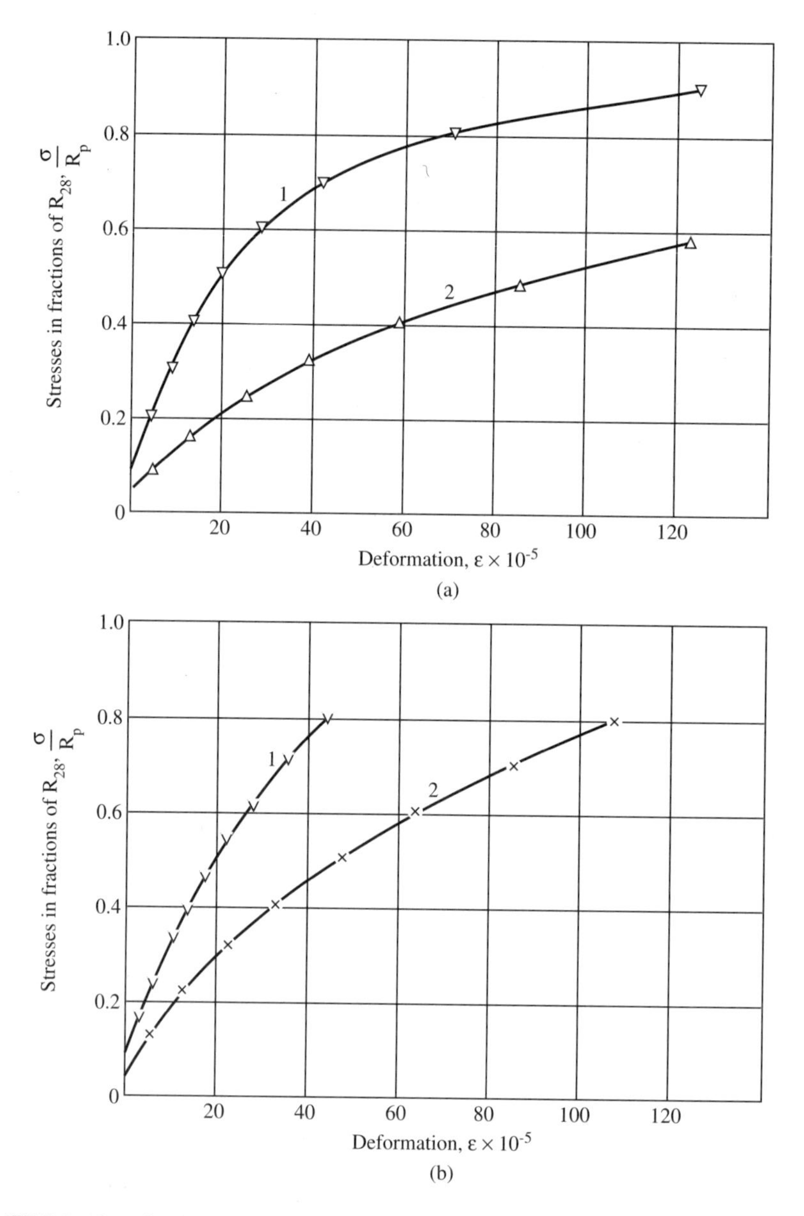

FIGURE 2.8 Longitudinal deformations vs. axial compression stresses of frozen and not frozen concrete. a — concrete after one day of curing before freezing, b — concrete after two days of curing before freezing, c — concrete after 28 days of hardening under normal conditions, d — frozen specimens, 1 — not frozen, 2 — frozen, 3 — concrete frozen immediately after placing in a form, 4 — model body, 5 — ice.

freezing was 12, 39, and 276 kg/cm^2. When tested, the specimens frozen after one and two days of precuring also were very plastic and were crumpled. The specimens frozen after 28 days of curing were characterized by brittle failure.

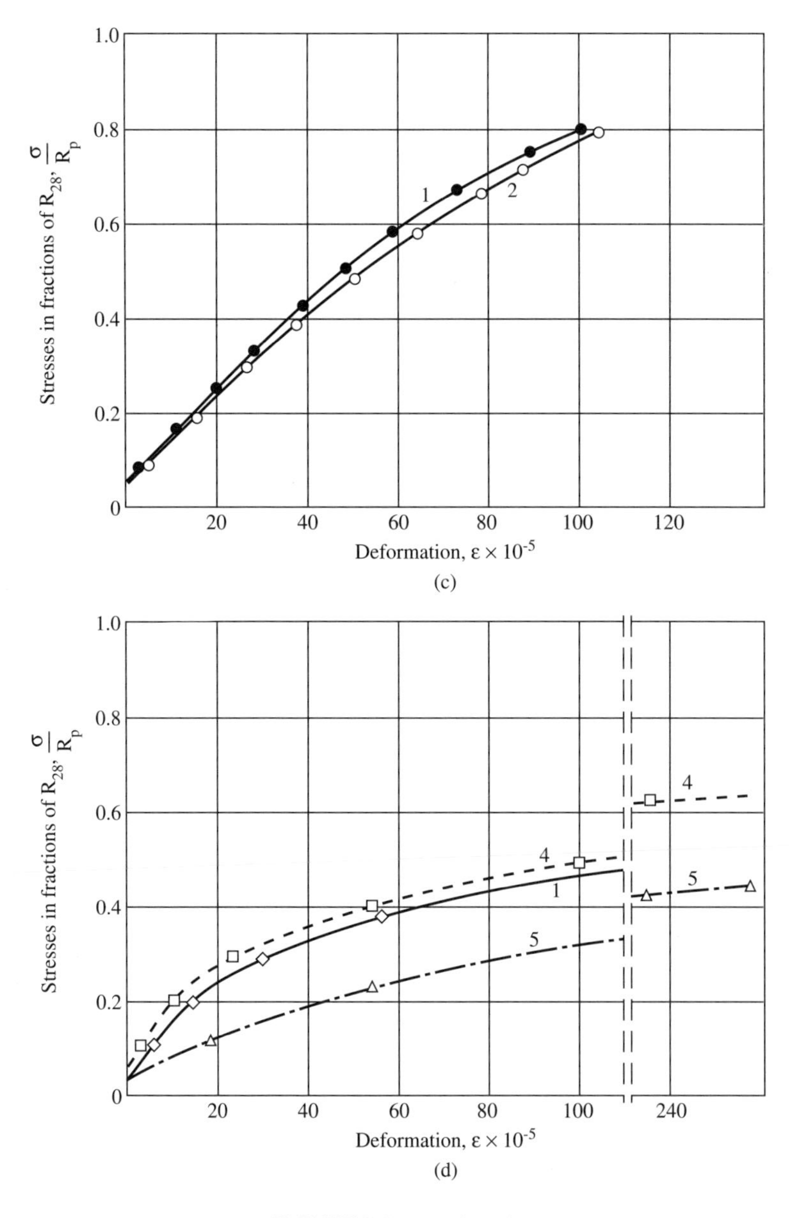

FIGURE 2.8 (continued)

Ice prisms also had brittle failure in testing and their strength was 58 kg/cm^2.

The bending-tensile strength of concrete found by testing beams of 100 × 100 × 400 mm was 77 kg/cm^2 in specimens frozen immediately after casting or 54% R of concrete that hardened for 28 days under normal conditions. The strength of the specimens that were precured for one, two and 28 days was 99, 116, and 176 kg/cm^2, respectively, or 69, 81, and 123% of the strength of concrete that hardened under normal conditions.

TABLE 2.4
Basic physico-mechanical properties of concrete in the frozen and unfrozen state

			Test results				Normative specifications	
Specimens	Conditions of freezing and thawing	Age of specimens	Cube strength	Prism strength	Bending-tensile strength	Modulus of elasticity × 10^3	Design compressive strength	Initial modulus of elasticity × 10^3
Frozen	Concrete frozen immediately after placing in mold	0	291/83	244/88	77/54	226/83		
	Concrete frozen after one day	1	407/116	365/132	99/69	250/92		
	Concrete frozen after two days	2	450/128	405/147	116/81	375/137		
	Concrete frozen after 28 days of hardening under normal conditions	28	511/146	450/166	176/123	386/141	350	295
	Ice	—	89	58	—	48.3		
	Model body	—	306	280	—	305		
Unfrozen	Concrete that hardened at 12–14°C	1	19/5.4	12/4.4	—	—		
	Concrete that hardened at 12–14°C	2	56/16	39/14.1	19/13	—		
	Concrete that hardened for 28 days under normal conditions	28	351/100	276/100	143/100	273/100	350	295

Note: The numerator shows absolute values and the denominator, values in % of R28 of concrete that hardened under normal conditions

The initial modulus of elasticity of concrete was determined on prisms of the same size as the ones used to find prism strength. The specimens frozen immediately after casting had a modulus of elasticity of 226 × 103 or 58% of the strength concrete that hardened under normal conditions. The modulus of elasticity of concrete that was cured before freezing for one day was a little higher — 250 × 10^3. The specimens frozen after two and 28 days of precuring had values of modulus of elasticity equal to 375 and 386 × 10^3, respectively, i.e., they exceeded the value of the concrete that hardened under normal conditions.

Creep deformation of concrete frozen at an early age also is of interest. An analysis of the experimental data (Figure 2.9) shows that loading of frozen specimens with the same initial loads of $0.2R_p$ produced quite different creeps in concrete of different ages by the time of freezing. The amount of ice in the frozen specimens affected to a great extent their values and nature. It can be seen particularly well in studying deformations of pure ice. As a permanent load was imposed on it for 20 days, ice deformations grew rapidly and its plastic properties were manifested distinctly. The reason must be the structure of the ice.

Ice crystal are hexagonal, i.e., they have the shape of hexagonal prisms. Each water molecule in the structure of ice is surrounded with four nearest molecules at the same distance from it. When water expands due to freezing, the closely packed structure is rapidly replaced by a loose crystalline structure that has many voids larger that the water molecules. This structure is unstable to effects of initial loading and rise in temperature. Under loading, the voids in the ice structure are filled with water molecules, the volume contracts and the density of a specimen reaches a maximum. Then the deformability decreases and smoothes out after 50 days of observations. Further observations found that the growth of deformations did not stop but their rate was much slower than at the beginning. The ultimate creep value characterizes flow plasticity of ice under loading.

Relative creeps of concrete specimens frozen immediately after casting were quite different both in the nature of variation and in values from those of ice. The quantity and quality of the concrete constituents was very important, of course, because ice functioned as a cementing substance. The resistance of specimens under a permanent load characterizes the adhesion between solid particles, i.e., the internal shear resistance. The thickness of the ice layer between solid particles and its pressure on the surfaces of the particles undoubtedly affected the deformability of frozen concretes.

Creep of immediately frozen specimens changed rapidly in the first 15 days whereupon it slowed down and increased monotonously at the end of observations. The role of cement was well identified in testing specimen models where cement had been replaced by quartz sand.

Observations revealed that relative creep deformations were the same in nature in those specimens but were 10% lower than those of concrete specimens. Their value appears to reflect the adhesion of the crushed quartz sand to ice in the frozen specimen models.

High creeps of specimen models and of freshly frozen concrete specimens took place because of the absence of a single solid skeleton. None of their particles were

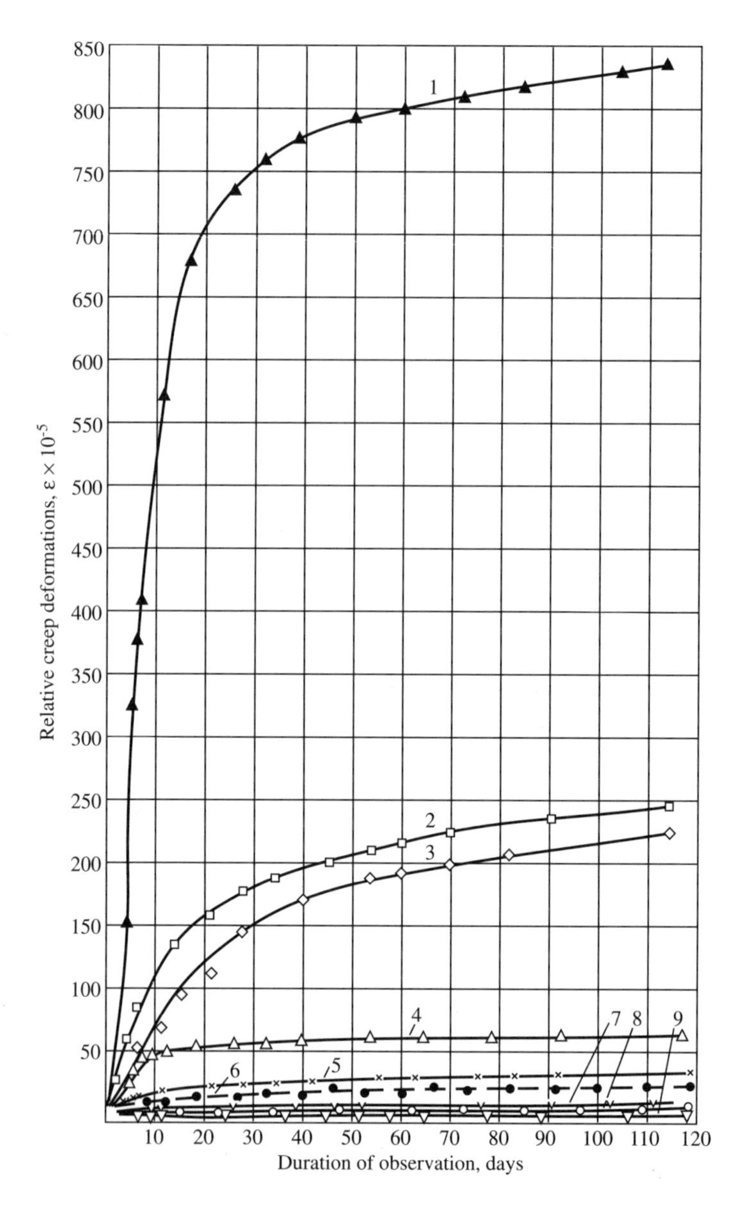

FIGURE 2.9 Relative deformations of frozen and not frozen concrete specimens under a load equal to 0.2 Rp. 1 — ice, 2 — concrete frozen immediately after placing in a form, 3 — model body, 4 — concrete frozen after one day, 5 — concrete frozen after two days, 6 — concrete frozen after 28 days of hardening under normal conditions, 7 — concrete loaded after 28 days of hardening under normal conditions, 8 — concrete loaded after two days of hardening under normal conditions, 9 — concrete loaded after one day of hardening under normal conditions.

connected into a single monolithic bound crystal system since they were divided by thin water films and ice crystals.

When a load was imposed, solid particles were displaced relative to each other due to high plasticity of the ice. That was the reason for the high deformations of these materials.

Relative creep deformations in specimen frozen after curing for one day were different in nature and rate. They sharply decreased and gradually stabilized after 18 days of observations.

Creep tended to decrease in specimens frozen after two days of precuring and after 28 days of storage under normal conditions, which can be seen quite well on the diagram. The reason was the formation of the structural crystal skeleton as the concrete hardened in precuring. These specimens contained much less ice than the previous freshly frozen ones. Cement had time to perform its function, which was to connect all the components into one solid body. The water became discontinuous and remained only in the free space between solid concrete particles. As the concrete hardened, the water that remained in it was mostly physically bound and lined the walls of air-filled pores and capillaries with monomolecular and polymolecular layers. When it changed into the solid aggregation state, the water expanded but failed to make solid particles displace in relation to each other because its expansion was largely compensated by compression of the air in the pores and capillaries. The presence of water films between crystals of hardened cement paste, however, was the cause of the increasing creep.

Concrete cured in a chamber at an above-zero temperature and not frozen had practically the same creep deformations as the specimens placed in the cooling chamber after one, two, and 28 days of precuring. The difference did not exceed 3%. These data agree quite well with the results of studies conducted by S. V. Alexandrovsky and V. V. Solomonov.

To sum up, we can draw certain conclusions about physico-mechanical properties of concrete in the frozen state. Basic properties of concretes in this condition are better as compared with those of concretes at the same age but hardened at an above-zero temperature. Their strength and other properties may considerably exceed those of check specimens with an increase in time of curing before freezing. This peculiarity of concrete in the frozen state should be taken into account in maintenance of reinforced concrete structures at subzero temperatures.

3 Preparations for Concreting in Winter Conditions

With rare exceptions, winter is an unfavorable season for construction work, concreting included. It involves some difficulties and makes building construction more expensive. To minimize the expected difficulties and to ensure high quality of concrete structures to be erected in winter with the least additional costs, the concreting in winter conditions should be carefully prepared in advance.

Prediction of temperature. The ambient temperature is a key factor because of its impact, on the productivity of the personnel in addition to kinetics of concrete hardening. So before the work can start it is necessary to get weather forecasts for a day, a week, and a month ahead from a meteorological service. Forecasts of daily temperatures enable builders to take timely precautions for concrete curing (to cover concrete members with additional layers of insulation or to heat them, and to estimate the time for form removal).

Weekly forecasts make it possible to select a proper concrete curing method well in advance and to prepare everything required for this purpose.

Monthly forecasts and weather forecasting for longer terms allow to take into account future temperature conditions in designs, to plan the operations correctly, to prepare the required equipment beforehand, and to provide comfortable conditions for the personnel.

The concreting can be carried out at any ambient air temperature but it becomes more difficult and costly as the temperature drops. For this reason, it makes sense to stop concreting operations at –30°C except for the Polar, Arctic, and Antarctic regions. Of course, when it is necessary to complete a construction project as soon as possible, the concreting can be continued at a lower temperature despite higher costs as compared with the summertime.

The choice of the concrete curing method. Curing of concrete after placing is the most important operation which should be performed competently and at a minimum cost.

The choice of the curing method is dictated by many factors, including the size, the function, the environment, and the reinforcement of a concrete structure, the ambient air temperature, and the construction schedule among others. The main thing is to select the most efficient method in technical and economic terms.

The existing methods used to cure concrete in the frost can be divided into two groups, including the ones that do not require an introduction of additional heat in the concrete after placing and those that always need heating in the course of curing. The first group comprises concretes that can harden in the frost (they contain

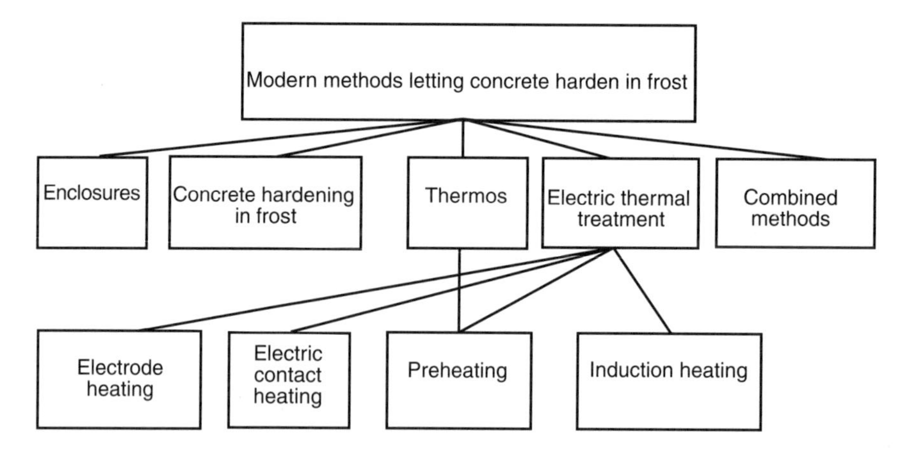

FIGURE 3.1 Concise classification of basic methods used to let concrete harden in the frost after placing.

antifreeze admixtures) and the thermos method; and the second group covers all methods that involve the heating of concrete (Figure 3.1).

The methods shown in the diagram, which are used to make concrete harden in the frost, are basic. Each of them has various modifications that cannot be discussed in detail in this book because of size limitations.

It should be emphasized that each method can help achieve optimum performance only in for certain applications that are the most suitable for it. None of the methods can be used efficiently to cure concrete in any structure and at any temperature. This should be kept in mind, and efforts should be made to find the most cost-effective method for a particular structure to be erected in the frost.

Once the curing method is set, it is necessary to prepare the materials and equipment needed to carry it out.

Unforeseen circumstances can arise, of course, during the work and these will need an on-line intervention. But the original decision is not changed, as a rule, but is only adjusted in some respects.

Preparation of the materials and equipment. Winter concreting always requires the materials that are not needed in the warm time of year. They include insulation materials to protect structures, chemical admixtures for concretes to harden in the frost, wire and metal rods for the electrodes used in electric heating, etc.

Very important is preparation of mixing materials if there is batching plant at the construction site. There is no need to store cement and aggregates for the whole winter period but the equipment required to thaw and heat aggregates and water should be provided in advance.

Water is heated in boilers or tanks with electric heaters. The capacity of a tank should correspond to the need in mixing water taking into account the heating temperature that may be as high as 40 to 80°C.

A convenient way of thawing and heating the aggregates is to use bins provided with electric heaters. It is uneconomical to heat aggregates to a high temperature because of high power consumption. To thaw and heat them to 8 or 10°C is enough.

Good practice is to produce a concrete mixture of the required temperature by heating the water. To prepare the concrete, aggregates and water are charged first into the mixer and cement is added after these have been thoroughly agitated.

Aggregates can be thawed and heated indoors in small stockpiles that have perforated pipes at the bottom to supply heated air to a stockpile. Steam is not recommended for this purpose. For one thing, it produces a lot of condensate in the aggregates, which is hard to estimate when you need a concrete mixture with a specified water–cement ratio and, for another, a warm wet aggregate may freeze solid at a low ambient air temperature because rooms where aggregates are thawed and heated usually have no heating. There have been many cases when aggregates heated with steam at a temperature well below zero froze and turned into a hard frozen mass when the supply of steam stopped for some reason. It was impossible to tear any aggregate away from that solid lump.

The delivery of mixed concrete to the placing site is a problem that needs careful consideration. If concrete is to be delivered from a mixer at the site, we can heat it to 60°C, deliver in closed bins, and place quickly in a structure. If the mixture is to be brought from a central batching plant and the interval between mixing and placing exceeds 15 min, it should be discharged from the plant with a temperature not higher than 30 or 35°C to avoid fast thickening. Concrete transportation facilities should be well insulated in a severe frost. When the work volume is high, concrete can be delivered along pipelines by a concrete pump. The pipelines should be insulated and flushed with warm water immediately after the work is completed. It is desirable that sodium nitrite be added to the water because it is an antifreezer and an inhibitor that can protect the pipe steel against corrosion.

Preparation of the formwork. The formwork used to cast structures in place in winter does not need any special precautions. Care should be taken to ensure reliable connections between formwork panels which must be made of metals that do not become brittle at a low subzero temperature.

However, if a reinforced concrete structure is to be erected in heating forms, the latter should be prepared in advance; electric heaters should be built in and securely fastened and insulated, the forms should be insulated from outside and have a protective cover to avoid damage and wetting.

Where metal or wooden forms will require lubrication, the lubricant should be selected so that it will not thicken too much at a low temperature. A water-base lubricant should be heated before it is applied to the surface of a form.

A serious problem that we often face in winter is cleaning forms from snow and ice before concrete can be placed in them. When there is an interval between these operations or in case of a sudden snowfall, snow may get inside a formwork already installed or the reinforcing bars may be covered with ice. An equipment should be provided for cleaning the reinforcement and forms before concrete is placed. The easiest and the most convenient procedure is to blow hot air from a hose connected to an air heater. Do not use steam to blow through the formwork because of water condensate that will be formed as a result. To prevent snow from getting inside the formwork, the latter should be always covered and the cover should be taken away only when concrete is to be placed.

Power supply. Winter concreting is impossible without a heat supply to the site. The heat is needed to keep concrete warm, to heat aggregates, and to provide warm shelters for the personnel. The heat supplied to the construction site is mostly of electric origin. Any estimation of requirements for electric power to drive machines and mechanisms should also include the above-listed uses. Provision should be made for application of electric power to heat rooms at a usual mains voltage, to preheat the concrete mixture at a voltage up to 380 V, or to heat placed concrete electrically at a voltage ranging from 36 to 120 V. Plan the electric power circuit well in advance at the design stage, including wiring to supply electricity to all the consumers.

A special emphasis should be laid on electric safety during operations to avoid fire or human injuries. Therefore, separate beds should be provided for electric preheating of concrete mixture (where only this method is used), guards should be installed for electric contact heating of placed concrete, and the equipment to turn on and cut off electric power should be installed in easily accessible places.

Instrumentation. Weather and climatic conditions, the temperature and humidity in the rooms for the personnel, and the temperature conditions while placed concrete is being cured must be monitored at all times in winter concreting. Instrumentation should be provided for this purpose, including thermostats for the rooms, which maintain a preset temperature automatically by switching on and off heaters; thermometers to measure the temperature of the concrete and of the outside air; electric regulators for concrete heating; panels to control the heating of concrete, with voltmeters and ammeters; hydrometers to measure the density of antifreezers when antifreeze admixtures are used; sclerometers; and ultrasonic apparatus and other devices for non-destructive tests of concrete.

Most of the above-listed equipment is kept at the on-job laboratory that monitors the curing of concrete and its strength development. On-job laboratories of large projects should have presses and other equipment to control the quality of cement, aggregates, admixtures, mixed concrete, and of the structures being erected. There is not much difference in organization of quality control laboratories for concreting in summer and in winter.

Comfortable conditions for the personnel. Comfort should be provided for engineers, technicians, and workers in the wintertime. Warm clothes and footgear should be procured, warm locker rooms equipped, and warming up and mess rooms provided. The amenity rooms are best to be made in insulated mobiles or assembled from modules whose set makes it possible to install and provide all amenity and locker rooms quite rapidly. The rooms largely use electric heaters. But sometimes, for construction projects in remote areas, wood, coal, gas, or steam supplied by steam generators are used for the purpose.

Special modules are installed directly on site to warm up workers when high rises or industrial installations are erected. As the storeys go up, the modules are moved up with them and so the workers do not have to leave their jobs and to waste time going down to get warm in a stationary shelter.

Attention should be paid to water supply for the locker rooms, showers, and mess rooms. The supply pipes should be insulated to prevent water from freezing and breaking them as a result. Waste water is either drained into a sewerage system if any or into a special insulated tank wherefrom it is daily pumped out and taken away by tanker trucks.

Documentation. All design and engineering documents related to winter concreting, including the concreting flow chart, the heating diagram with the layout of electric equipment, the curing temperature log, winter concreting manuals, the layout of holes for temperature measurements in the structure, etc. should be completed by the time the winter season starts.

All matters of reinforcing and formwork operations and of mixing, placing and curing of concrete should be worked out in detail. Emergency measures should be planned in case of a sudden drop in the outside air temperature or a heavy snowfall so that proper conditions could be provided for concreting and curing at all times.

It should be emphasized in conclusion that preparations for concreting in winter are an important stage in construction and the pace of erection of buildings and structure and the quality of reinforced concrete members depend on them in no small degree.

4 Concretes Containing Antifreeze Admixtures

The methods that let concrete harden at a subzero ambient temperature without applied heating can be divided into two groups: use of concretes containing chemical antifreeze admixtures, which can harden at subzero temperatures, and the thermos method that lets placed concrete harden at an above-zero temperature.

The methods that do not use applied heating in winter concreting are the most economical ones since they do not need energy resources for curing until concrete develops the required strength. This is the reason why they are adopted in all cases where these methods are feasible technologically and economically.

The antifreeze-containing concrete that can harden in the frost consists of ordinary constituents but mixed with a water solution of a chemical agent that freezes at a temperature well below 0°C.

The chemical agent becomes an inherent component of the concrete in the form of an admixture. Such an agent can impart certain properties to the concrete mixture as it is handled, affect the process of hardening, and finally ensure the required properties of the concrete during the service life of the concrete structure. When buildings and engineering structures are erected in winter, the most important phase is the curing of concrete after it was placed and until it attains the required strength. The antifreeze admixtures (antifreezers) used for this purpose must meet certain specifications. They should:

- make concrete harden at a subzero temperature without applied heating;
- ensure that the structure and properties of the concrete meet standard specifications;
- not be corrosive to reinforcing steel;
- not be toxic;
- not be in short supply and expensive.

There are many natural chemical substances whose water solutions freeze at temperatures lower than 0°C. However, just a few of the whole list were found to be acceptable for use in concrete. These include sodium nitrite ($NaNO_2$), potassium carbonate (K_2CO_3), chlorides –$CaCl_2$ and NaCl–, and some composites that combine several admixtures.

4.1 THE PRINCIPLE OF THE METHOD AND ITS THEORY

The idea of the method is as follows. Concrete mixed with a solution of an antifreeze admixture continues to harden below zero because of the presence of water in the liquid phase.

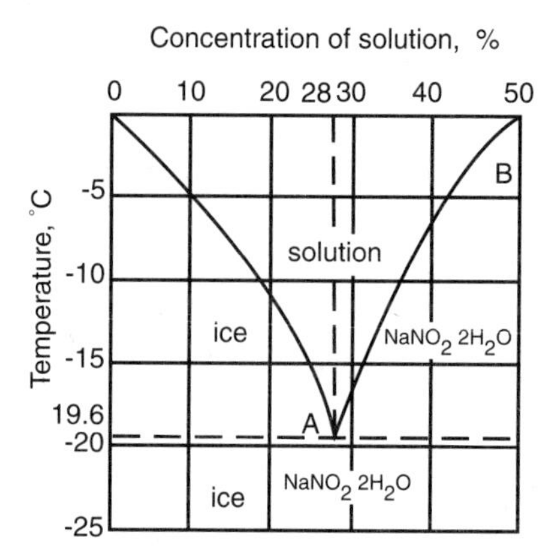

FIGURE 4.1 A diagram of state of water solutions of sodium nitrite at temperatures below 0°C.

The freezing point of the water is depressed due to reaction with the antifreezer when it is dissolved. Solvates thus formed are more or less strong compounds of particles of the dissolved substance with water molecules, e.g., Na^+ and NO_2^-; ions when sodium nitrite is dissolved. The conversion of the water solution to ice will require energy to be expended not only to slow down the motion of water molecules but also to break the solvates. Their composition is characterized by the number of water molecules bound with each particle of the dissolved substance. The strength of the bond depends mainly on the electric properties, size and combinations of the particles and on the content of the particles in the water unit volume, i.e., on the concentration of a solution. When the concentration increases, the amount of bound water grows and the freezing point of the solution drops. However, the quantity of free water molecules that can react with the minerals of the cement clinker decreases.

The formation of the solvates causes the water to freeze gradually in the solution as it cools down. For example, when a sodium nitrite solution with a concentration lower than 28% cools down, "fresh water" ice crystals start precipitating from it as soon as the temperature of the solution reaches a value in the curve OA (Figure 4.1). From this moment on, the fresh water ice crystals continue to precipitate as the cooling continues while the concentration of the solution increases along the same curve OA. Upon reaching the eutectic point A, where the concentration is 28% and the temperature is –19.6°C, the remaining water changes completely to the solid phase, a mixture of ice and crystals of $NaNO_2 \times 2H_2O$.

If a solution of sodium nitrite with a concentration higher than 28% is cooled down, the concentration, on the contrary, decreases due to the precipitation of the $NaNO_2 \times 2H_2O$ crystals. The concentration changes along the curve BA, the solution freezes again at the eutectic point A, and ice and crystals of $NaNO_2 \times 2H_2O$ are formed. The eutectic point is constant for the solution of each admixture and does not depend on its initial concentration. The latter affects the amount of ice or of excessive substance crystals that will precipitate from the solution before the eutectic point is reached and the quantity of ice that will form when the solution freezes completely.

Thus, the eutectic point is the limit to which the freezing temperature of a water solution of an antifreeze admixture can drop. In building practice, the temperature limit of application of any antifreezer is generally set higher than its eutectic point. So it is recommended that sodium nitrite be used for a temperature of concrete not lower than –15°C while its eutectic point is –19.6°C. As was mentioned above, the reason is that a higher concentration of the solution reduces its content of the free water molecules that can react with the minerals of the cement clinker.

Another factor is that antifreeze admixtures are added to concretes in quantities less than required to form a completely unfrozen liquid phase at a particular designed hardening temperature. The conversion of 30 to 50% of mixing water to ice when the antifreezer containing concrete cools down to the design temperature does not adversely affect its properties while a the concentration of a solution higher than the optimum value retards the hardening process.

The introduction of antifreeze solutions into a concrete mixture greatly affects the ice that is formed in freezing. It has a faulty, scale structure, low strength, and cannot cause any noticeable damage to concrete. In contrast to this, ice formed in concrete without any antifreeze admixture can reduce considerably its strength characteristics and durability. So, due to the above features of the ice formed in the antifreezer-containing concrete, the critical strength that allows concrete to freeze can be lower than that of concretes without antifreeze admixtures. The building code of Russia, for example, has established the following critical strength values for antifreezer-containing concretes:

- 30% for concretes with R_{28} up to 200 kg/cm^2;
- 25% for concretes with R_{28} up to 300 kg/cm^2; and
- 20% for concretes with R_{28} up to 400 kg/cm^2 or more.

An antifreezer added to a concrete mixture, along with depressing the freezing point of water, also participates in the cement hydration processes and changes the solubility of the cement clinker minerals and of the resultant hydrates. The substances used as antifreeze admixtures affect the properties of water in different fashions and participate in the processes of the hydration of cement and of the hardening of concrete in different ways.

If we compare the strength development rate of concrete without any admixture at 20°C and that of the same concrete but containing an antifreezer at a subzero temperature, we will see that the latter hardens two or three times slower. Having benefited from saving energy resources by using concrete that can harden in the frost we lose in the hardening rate. So the application of this method in winter concreting can be clearly defined: antifreezer-containing concretes can be used in structures that do not need to be stripped of the formwork in the shortest time possible, i.e., where the construction schedule is not tight.

Antifreeze admixtures can be used in structural concretes with dense or porous aggregates. However, you should know specific features of each admixture quite well to be able to use it to advantage and to produce a high-quality concrete structure.

Calcium and sodium chlorides were the first to be used as antifreeze admixtures early in the 1950s. Chlorides are quite strong antifreezers. The eutectic point of the solution of calcium chloride is at the level of –55°C and that of sodium chloride at the level of –21°C. So properly concentrated solutions of these salts will not freeze at a temperature higher than these points thus providing enough water in the liquid phase to react with cement. Despite warnings from researchers, builders started to use these admixtures not only in plain concrete but in reinforced concrete structures on a large scale. The aggressivity of chlorine toward reinforcing steel is well known and it manifested itself in a couple of years. Hairline cracks appeared and rapidly widened in the concrete cover along the reinforcement. The reason was the reaction of chlorine with iron. The reaction products were larger in volume than the reactants. The rust layer on the steel surface gradually increased, brought pressure to bear on the concrete, and tensile stresses arose in the concrete cover. Since the resistance of concrete to tension is eight to 10 times lower than to compression, these stresses caused the hairline cracks in the cover when they reached a critical value.

A hairline crack in the cover made steel accessible for oxygen and water vapor from the air. This sharply accelerated corrosion and the layer of rust on steel bars started to grow fast thus widening the crack and letting in more corrosive agents from the environment. Later on the cover may even fall off entirely and the structure may fail.

For the above considerations, the Research Institute of Concrete and Reinforced Concrete prohibited the use of chlorides as antifreeze admixtures to concretes in reinforced concrete structures. Calcium chloride may be used in the latter only as a hardening accelerator and its quantity must not exceed 2% by cement weight, except for prestressed structures where it cannot be used in any capacity.

Chlorine-free admixtures are used for winter concreting. The most common are sodium nitrite ($NaNO_2$) and potassium carbonate or potash (K_2CO_3). These anti-freezers do not cause corrosion of steel and let concrete freeze at a temperature from –15°C ($NaNO_2$) to –25°C (K_2CO_3). Each has its own peculiarities and their actions on concrete are different.

Addition of sodium nitrite. Sodium nitrite belongs to the group of antifreeze admixtures that depress the freezing point of the liquid phase in concrete and practically do not affect the structuring rate. The admixture lets the concrete harden in the frost mainly because it keeps the liquid phase unfrozen.

Sodium nitrite is not a strong antifreezer. It does not react directly with the minerals of the cement clinker and participates only in exchange reactions. An important advantage of sodium nitrite is its inhibiting property that enables it to protect steel against corrosion rather than cause it. For this reason it can be widely used in reinforced concrete structures.

Concrete mixtures containing sodium nitrite are somewhat more plastic (not much) and have the same setting time as concretes without the admixture. Such mixtures are very easy to handle.

The quantity of sodium nitrite to be added to concrete vary according to temperature. It is 4 to 6% when the temperature of concrete is no lower than –5°C, 6 to 8% for any temperature between –6 and –10°C, and 8 to 10% of the dry cement weight at a temperature from –11 to –15°C. Moisture content of aggregates should be taken into account because the concentration of the sodium nitrite solution is usually 20%.

TABLE 4.1
The development of strength by Portland cement concrete containing antifreeze admixtures

Admixture	Design temperature in°C	Strength in % R_{28} developed in concrete hardening in frost during following period in days			
		7	14	28	90
Sodium nitrite	–5	30	50	70	90
	–10	20	35	55	70
	–15	10	25	35	50
Potassium carbonate	–5	50	65	75	100
	–10	30	50	70	90
	–15	25	40	65	80
	–20	25	40	55	70
	–25	20	30	50	60

Concrete containing sodium nitrite develops strength rather slowly (Table 4.1).

Addition of potassium carbonate. Potassium carbonate is a strong antifreezer (its eutectic point is at the level of –36.5°C) that accelerates the hardening of concrete at an early age. The reason for the acceleration is that this admixture changes the solubility of the silicate components of cement and forms double salts with hydration products.

When it reacts with the hydration products of aluminum containing phases (C_3A and C_4AF), potassium carbonate forms calcium hydrocarboaluminate. Its reaction with calcium hydroxide that evolves during the hydration of cement produces hydrocarbonate and calcium carbonate (accumulating simultaneously caustic potassium with an eutectic point of –36°C in the liquid phase of the concrete). These new phases participate in the formation of the structure of hardened cement paste. The crystals of the salts thus formed are usually elongated and so they reinforce as it were the cement paste.

Since potassium carbonate takes an active part in the hydration processes, its optimum quantity for a given subzero temperature of concrete depends to a great extent on the chemical, mineralogical, and material composition of cement. So it becomes necessary always to check the concrete hardening rate and to determine accurately the proportioning of the antifreezer and the concrete setting time with the cement to be used for temperatures at which the concrete mixture will be handled.

If the setting time is short, it is advisable that a retarder be added to the concrete mixture to avoid its rapid thickening. It may be sulfite-yeast spent grains, sodium tetraborate (borax) with the formula $Na_2B_4O_7 \times 10H_20$, or a two-component admixture comprising water glass (soluble sodium silicate) and sodium adipinate (sodium salt of adipic acid) in 1:1 ratio. Other setting time retarders also may be used.

The quantity of a retarder to be added to concrete depends on that of potash and may be approximately as shown in Table 4.2.

The numbers shown in Table 4.2 are approximate, and the exact quantity of a retarder should be determined by the on-job laboratory.

TABLE 4.2
Recommended quantities of setting time retarders for concretes containing potassium carbonate

Proportioning of K_2CO_3 in % of cement weight	Quantity of setting time retarders in % of cement weight		
	Sulfite-yeast spent grains	Sodium tetraborate	Water glass + sodium salt of adipic acid
5–6	0.5–0.75	1.0–1.2	0.8–1.2
6–8	0.5–1.0	1.2–1.6	1.0–1.6
8–10	0.75–1.0	1.6–2.0	1.2–2.0
10–12	1.0–1.25	2.0–2.4	1.6–2.6
12–15	1.0–1.25	2.4–3.0	1.8–3.2

The setting time of concrete containing potassium carbonate may be increased by using certain special procedures. One procedure is to prepare concrete by mixing first its constituents thoroughly with water and only then adding a concentrated solution of potassium carbonate. The total content of the water used for the mixing and of the water in the concentrated solution of potassium carbonate should not exceed that required for the water–cement ratio selected for the given concrete. In this case, when concrete constituents are mixed with water, solvate films are formed around cement grains. They prevent the direct contact of potassium carbonate with the cement grains for about half an hour. Later on, as the solvate films gradually disappear, the reaction of potassium carbonate with the concrete takes its usual course.

The second procedure is to mix cement and water with cold aggregates that have a subzero temperature making sure that the latter are not covered with ice, frozen into lumps, or have inclusions of snow and ice. Then add a potassium carbonate solution to the mixture and the admixture will react with the cement clinker minerals slowly due to a low reactivity of the water in the solution. Thus, the setting time will be longer.

The quantity of potassium carbonate to be added to concrete depends on the temperature at which the concrete is going to harden. It is 5 to 6% of the dry cement weight for a temperature no lower than –5°C, 6 to 8% when the temperature is between –6 and –10°C, 8 to 10% for temperatures from –11 to –15°C, 10 to 12% for temperatures from –16 to –20°C, and 12 to 15% for temperatures between –21 and –25°C.

The potassium carbonate admixture may migrate in the course of curing, accumulate in certain areas of a structure, such as ribs or surface layers, and crystallize. The most favorable conditions for this process are created by the appearance of temperature moisture gradients that accelerate mass transfer. The causes are repeated temperature drops and rises over zero point when the structure cools down and is warmed up many times in autumn or spring, or during thaws in winter. Since the crystallization of salts increases their volume, their accumulations in certain areas may produce structural defects in the concrete there.

TABLE 4.3
Curing time of antifreezer-containing (K_2CO_3) concrete until it develops critical strength

Admixture	Designed concrete temperature in °C	Curing time in days for concrete grade		
		200	300	400
Potassium carbonate	–5	3	2	1
	–10	7	5	4
	–15	9	7	5
	–20	9	7	5
	–25	14	10	7

The double salts — hydrates — thus formed are potentially dangerous components of the hardened cement paste when the antifreezer-containing' concrete is used in a corrosive water medium. This is the reason why their use is restricted for such environments.

Concrete containing potassium carbonate develops its strength in the frost a little faster than that with sodium nitrite (Tables 4.3 and 4.4).

Antifreezer-containing concrete mixed with cement that has a lot of mineral additives, e.g., Portland pozzolana cement, hardens slower, particularly at a temperature lower than –10°C, and its strength development rate is lower than the values given in Tables 4.3 and 4.4.

4.2 SELECTION AND ASSIGNMENT OF ANTIFREEZE ADMIXTURES

An antifreeze admixture is selected according to the design temperature at which concrete is supposed to harden. If the temperature rises above the design value or even becomes positive as the concrete is cured, its hardening will accelerate and it will attain the design strength earlier than planned.

Concrete containing the quantity of an antifreeze admixture designed for a certain hardening temperature should not be subjected to a lower subzero temperature to avoid freezing. The latter may adversely affect its properties after thawing due to irreversible deterioration of the material's structure under the action of frost. The concrete must reach its critical strength by the time of freezing whereupon it can be frozen without damage to its structure and properties (Tables 4.3 and 4.4). The only thing that has to be put up with in this case is temporary retardation or even stoppage of hardening if the temperature drops below the design value. As the temperature rises and the concrete thaws, the hardening starts again, and if the temperature rises well above the design value, the hardening accelerates even more and the design strength may be attained earlier than shown in Tables 4.3 4.4.

The selection of an antifreeze admixture must be closely coordinated with the composition of concrete, the type of structure, and its environment.

TABLE 4.4
Curing time of antifreezer-containing ($NaNO_2$) concrete until it develops critical strength

Admixture	Designed concrete temperature in °C	Curing time in days for concrete grade		
		200	300	400
Sodium nitrite	–5	7	6	4
	–10	12	9	7
	–15	19	14	11

Sodium nitrite may be used in concrete for almost any reinforced concrete structure, but there still some restrictions. This admixture cannot be used:

- in concretes for prestressed structures reinforced with steel of classes At-IV, At-V, At-VI, and A-V;
- in reinforced concrete structures and joints of precast or cast-in-place structural members that have protruding reinforcement bars or inserts of aluminum-plated steel;
- in reinforced concrete structures designed for service in water or gaseous media with a relative humidity over 60% if the aggregate contains inclusions of reactive silica;
- in reinforced concrete structures for electric transportation facilities and For industrial enterprises that use direct current.

It can be seen from the list of the restrictions that they do not narrow much the applications of sodium nitrite in reinforced concrete structures.

Potassium carbonate is more restricted in use for concrete and reinforced concrete structures. In addition to the above-listed cases that are not recommended for sodium nitrite, it cannot be used in:

- prestressed structures, in joints with prestressed tendons, and in the channels for the tendons;
- in reinforced concrete structures and joints of precast or cast-in-place structural members that have protruding reinforcement bars or inserts of zinc-plated steel;
- in reinforced concrete structures designed for operation in corrosive sulfate waters, in solutions of salts or caustic alkalis when there are evaporating surfaces, and in zones of variable water level.

This list of the cases where potassium carbonate cannot be used, although longer than that for sodium nitrite, still leaves wide room for applications in reinforced concrete structures.

A problem that is often faced in practice is the appearance of efflorescence (traces of salt) on the concrete surface during service. There are many reasons for it and addition of admixtures to concrete is just one of them. These may be not only antifreezers but also other agents added to impart certain properties to concrete. The effloreseence may be caused by some components of cement, aggregates, or their impurities and it would be wrong to blame the admixtures in this case.

The procedure described in the *Manual for Use of Chemical Admixtures* published in Russia can be used to check whether effiorescence can appear on the surface of a particular concrete. It consists in the following:

Make three specimens — prisms of 70 × 70 × 280 mm or 100 × 100 × 400 mm — of concrete with a permissible minimum, medium, and maximum amount of the admixture intended for use in the concrete and three similar specimens of concrete without the admixture. After the specimens were cured under the conditions prescribed for the given type of operations immerse the prisms in water to a depth of three to five cm. Each series of specimens should be placed in a separate container. Blow air at a temperature of 20 to 30°C over the exposed surface of the specimens.

Inspect the specimens regularly during the test. The presence of efflorescence can be seen visually by the appearance of bleached spots or salt traces. The absence of efflorescence for seven days shows that the admixture can be used in the concrete on whose surface it is unacceptable.

Antifreezers can be used in combination with other admixtures, such as air-entraining agents, hardening accelerators or retarders, and plasticizers. Superplasticizcrs are particularly advisable for use with antifreeze admixtures. Studies found that the addition of a superplasticizer together with an antifreezer to a concrete mixture can reduce the amount of the latter nearly by half as compared with that designed for the given predicted temperature of hardening. And the concrete hardens faster and attains the design strength in one or two months.

It is good practice to add an antifreezer to concrete that is to be heated after placing by any method. In a severe frost, a small amount of an antifreeze admixture (2 to 4% by cement weight depending on the ambient air temperature) can protect concrete against some freezing during transportation and placing before the heating is started. If the antifreezer-containing concrete is to be heated with electrodes or electrically preheated, its specific resistance must be checked (it will be lower than that without the admixture) and appropriate adjustments should be made in the layout of the electrodes and in the selection of electric voltage.

The thermal treatment of antifreezer-containing concrete can produce the stripping strength of the concrete in a very short time. The combination of the admixture with heating is particularly effective for concreting in a severe frost when the air temperature drops below –25°C and when antifreeze admixtures alone cannot make the concrete harden at such temperatures. This is especially true for electric contact heating of concrete. At a very low temperature, placed concrete may freeze quickly as it touches a cold form and electrodes attached to it and the electric heating will become impossible because frozen concrete has a high resistance and the applied voltage will be insufficient to overcome it. An antifreeze admixture depresses the freezing point of the concrete retarding the freezing process and, moreover, reduces the specific resistance of the concrete. In this case we can start the electric heating even if the concrete cooled down below 0°C.

The electric heating of concrete containing an antifreeze admixture makes it possible, under optimum conditions, to produce a compressive strength of 75 to 90% by the time it is completed. Then the strength of the concrete may increase to 100 or 120% of its design value after 28 days of curing in the frost and 28 days of curing in normal temperature and humidity conditions.

If there is no need to attain this high strength by heating, the thermal treatment time can be reduced considerably at the isothermal curing stage. It s electric power and increases the turnover of forms while the quality of the concrete remains quite high without sacrificing its basic properties, including tensile strength, the bond with the reinforcement, and frost resistance, as compared with concrete that hardened without electric heating.

Sodium nitrite is recommended as an antifreeze admixture to be added to the electrically heated concrete. Potassium carbonate is inadvisable for such cases because concretes heated with this admixture do not develop strength as planned. Sometimes it may be lower than the required level by as much as 30%. The frost resistance and water impermeability of such concrete also are low.

Concrete with low aluminate Portland cement containing up to 6% C_3A in the clinker is preferable for combining electric heating with an antifreeze admixture to avoid the underdevelopment of strength.

Electric heating of concrete containing an antifreeze admixture should be preceded by laboratory tests of specimens made of the same concrete that is to be placed in a structure.After they are frozen to a subzero temperature, the specimens are electrically heated to determine their strength afterwards as compared with that of a reference specimen.

4.3 PROCEDURES

Procedures of concreting with antifreeze admixtures do not differ much from those for ordinary concretes. Special emphasis should be laid on preparation of the water solution of an antifreezer and on concrete mixing.

Admixtures are generally added to concrete in the form of a water solution of working concentration. This allows to proportion admixtures correctly and to spread them evenly over the concrete mass. The admixture solution of working concentration should be such that the concrete will not need any additional mixing water. The solution can be prepared either in advance, for which purpose it needs additional tanks, or in a water proportioning plant.

In the first case, the tanks for preparing admixture solutions are provided with a system of pipes for agitation with compressed air and with heaters if necessary. The solution of working concentration flows from the preparation tank into a discharge hopper wherefrom it is fed to mixers via the water proportioning plant. The hopper has level gages with automatic control of filling. Special care should be taken to proportion correctly the components in the preparation tank by measuring the density of the solution being prepared. To obtain the solution of the required concentration from a solid product, the latter is dissolved on the basis of a certain dry content in 1 liter of water as shown in Table 4.5.

TABLE 4.5
The content of solid admixtures for preparation of their water solutions

Required concentration of solution in %	Content of dry product in kg per 1 liter of water	Required concentration of solution in %	Content of dry product in kg per 1 liter of water	Required concentration of solution in %	Content of dry product in kg per 1 liter of water
2	0.020	16	0.190	30	0.429
4	0.042	18	0.220	32	0.470
6	0.064	20	0.250	34	0.515
8	0.087	22	0.282	36	0.563
10	0.111	24	0.316	38	0.613
12	0.136	26	0.351	40	0.667
14	0.163	28	0.391	42	0.721

When the solution of working concentration is prepared in a water proportioning plant, the admixture is delivered from storage to the preparation tank where a highly concentrated solution is prepared. The tank is provided with pipes to agitate the solution with compressed air and with an automatic temperature control system whose accuracy is ±2°C. The concentrated solution is flows to a discharge hopper wherefrom it is supplied to the water proportioning plant through a liquid proportioning plant. Water is fed to the plant to dilute the solution to a working concentration whereafter it goes to the concrete mixer. It is recommended that the concentration of an antifreeze admixture in this case should be 20 to 30%.

The superplasticizer also enters the water proportioning plant together with the antifreeze admixture in a highly concentrated form. As was mentioned above, when a superplasticizer is added to concrete with an antifreeze admixture, the quantity of the latter may be reduced by half for the same ambient air temperature.

Antifreeze admixtures may come as highly concentrated liquids as well as solids. The solution of working concentration is then prepared by adding water until the required density is obtained. The density is a measure used to monitor the concentration.

When the solution is prepared from a solid antifreeze admixture, the water is better be heated to 40 or 80°C and agitated to speed up dissolution. Hard lumps, if any, should be crushed before dissolving.

The solutions of antifreeze admixtures are prepared in carefully washed tanks protected against rain or snow. The admixture solution should be thoroughly agitated before use because solutions with sediments of undissolved components arc prohibited. The solutions may be stored at subzero temperatures whose permissible values depend on their concentrations (Tables 4.6 and 4.7).

The mixing procedure for concrete containing an antifreeze admixture does not differ from that for concrete without the admixture if the temperature of its aggregates is above zero. If the aggregates have a subzero temperature, they should be charged in the mixer first together with the admixture solution of working concentration. After these were stirred for 1.5 or two minutes, cement is added and the mixture is

TABLC 4.6
The sodium nitrite content, density, and freezing point of solutions

Concentration of solution (%)	Solution density at 20°C in g/cm³	Content of dry sodium nitrite in Kg		Freezing point in°C
		in 1 L solution	in 1 kg solution	
2	1.011	0.020	0.02	−0.8
4	1.024	0.041	0.04	−1.8
6	1.038	0.062	0.06	−2.8
8	1.052	0.084	0.08	−3.9
10	1.065	0.106	0.10	−4.7
12	1.078	0.129	0.12	−5.8
14	1.092	0.153	0.14	−6.9
15	1.099	0.164	0.15	−7.5
16	1.107	0.177	0.16	−8.1
17	1.114	0.189	0.17	−8.7
18	1.122	0.202	0.18	−9.2
19	1.129	0.214	0.19	−10.0
20	1.137	0.227	0.20	−10.8
21	1.145	0.240	0.21	−11.7
22	1.153	0.254	0.22	−12.5
23	1.161	0.267	0.23	−13.9
24	1.168	0.280	0.24	−14.4
25	1.176	0.293	0.25	−15.7
26	1.183	0.308	0.26	−17.0
27	1.191	0.322	0.27	−18.3
28	1.198	0.336	0.28	−19.6
29	1.206	0.350	0.29	(−17.8)
30	1.214	0.364	0.30	(−16.5)
32	1.230	0.394	0.32	(−14.0)
34	1.247	0.424	0.34	(−11.7)
36	1.264	0.455	0.36	(−9.5)
38	1.282	0.488	0.38	(−7.5)
40	1.299	0.520	0.40	(−6.0)

Note: The brackets contain the temperature at which excessive salts precipitate before thc solution freezes at the eutectic point, which is −19.6°C for water solutions of sodium nitrite.

agitated for another four or five minutes. The temperature of concrete containing sodium nitrite as it is discharged may be 15 to 35°C. The temperature of concrete with potassium carbonate must be 15°C or lower so that it could drop below zero during the setting and initial hardening of the concrete.

Concrete mixtures with the above temperatures have optimum setting times and conditions for the formation of cement hardened paste. They do not harden rapidly enough at the beginning. Where the temperature of a concrete mixture is lower than required, measures should be taken to have its temperature exceed the freezing point of the mixing solution by at least 5°C after placing and compaction. The required temperature of a concrete mixture is set by the on-job laboratory that takes into

TAB1E 4.7
The potassium carbonate content, density and the freezing point of solutions

Concentration of solution (%)	Solution density at 20°C in g/cm³	Content of dry potassium carboiiate in kg: in 1 L of solution	Content of dry potassium carboiiate in kg: in 1 kg of solution	Freezing point in 0°C
2	1.016	0.020	0.02	–0.7
4	1.035	0.041	0.04	–1.3
6	1.053	0.063	0.06	–2.0
8	1.072	0.086	0.08	–2.8
10	1.090	0.109	0.10	–3.6
12	1.110	0.133	0.12	–4.4
14	1.129	0.158	0.14	–5.4
15	1.139	0.171	0.15	–5.9
16	1.149	0.184	0.16	–6.4
17	1.159	0.197	0.17	–7.0
18	1.169	0.210	0.18	–7.6
19	1.179	0.224	0.19	–8.2
20	1.190	0.238	0.20	–8.9
21	1.200	0.252	0.21	–9.6
22	1.211	0.266	0.22	–10.3
23	1.221	0.281	0.23	–11.2
24	1.232	0.296	0.24	–12.1
25	1.243	0.311	0.25	–13.0
26	1.254	0.326	0.26	–14.1
27	1.265	0.341	0.27	–15.1
28	1.276	0.357	0.28	–16.2
29	1.287	0.373	0.29	–17.4
30	1.298	0.390	0.30	–18.7
32	1.321	0.423	0.32	–21.5
34	1.344	0.457	0.34	–24.8
36	1.367	0.492	0.36	–28.5
38	1.390	0.528	0.38	–32.5
40	1.414	0.566	0.40	–36.5

account its thickening time; heat loss during transportation, handling, and placing, and cost effectiveness.

Concertes to be placed in joints between precuts units in small quantities, to be cast on a structure of steel shapes, or to be placed on a frozen base and with a condensed reinforcement should be mixed at a temperature that takes into account heat losses for warming the steel, forms, etc. to avoid freezing of concrete directly after placing. The temperature of placed concrete in these cases should be at least 5°C higher than the freezing point of the mixing solution.

Concrete containing an antifreeze admixture can be transported in a container without insulation but it should be always protected against atmospheric precipitation or freezing of water. The concrete should have the specified temperature and workability when it arrives at the placing site. If these requirements cannot be met, the concrete should be transported in an insulated and closed container.

As always, the antifreezer-containing concrete should be placed in a structure cleaned of snow and ice, the exposed surfaces being covered to protect it against snowfalls and additional heat losses. When there is a snowfall or strong wind, concrcte with or without an admixture is placed in a tarpaulin shelter or a light enclosure.

The hardening of placed concrete should be closely monitored. If its temperature drops below a design value, the structure is covered or even heated until the critical strength is attained by the time of freezing.

5 Thermos Curing

5.1 THE PRINCIPLE AND APPLICATION OF THE THERMOS METHOD

The thermos method is one of the most cost-effective techniques used to cure concrete structures cast in place in winter. The idea of the method is to let concrete harden by maintaining an above-zero temperature in the concreted structure without additional applied heating. The temperature is maintained by the heat that was introduced into the concrete in mixing plus the heat evolved during hydration of cement. The method is simple and the curing of concrete in this fashion does not differ much from the summertime. So it requires less additional expenditures than other methods of curing concrete in cast-in-place structures erected at a subzero ambient temperature.

The feasibility and suitability of the thermos curing for winter concreting depend on various factors, where the most important are (1) the size of a structure, (2) the reactivity of cement, (3) the temperature of concrete to be placed, (4) the ambient air temperature, (5) wind velocity, and (6) the period during which the project is to be completed. A combination of the key factors governs the ability of the method to let the concrete attain the required strength by the time the structure cools down and the concrete freezes. It means that the concrete may develop by that time the design strength (a rare occasion), the form striking strength as specified for winter, or the critical strength after which it may freeze without damage.

The specific features of the method dictate its applications. It is recommended that the thermos curing be used to cast in place massive structures with a surface modulus up to 5 $\left(M = \frac{\Sigma F}{V}\right)$. The surface modulus, which is a ratio of the sum of areas of a structure's surfaces exposed to cooling in sq.m to its volume in cu.m, should be calculated before a decision to adopt this method is made. The surface modulus is given for some structures in Table 5.1. It is assumed that the structure's surface that contacts the thawed base is not considered to be a surface exposed to cooling.

When the thermos curing is used, placed concrete is well insulated from the formwork side, and its exposed surface is covered.

Design monthly temperatures of the ambient air and the wind velocity for the region where a structure is to be cast are taken into account in planning winter concreting operations and estimating the capabilities of the thermos method. For example, the design monthly temperatures in Moscow are –8°C in November, –19.1°C in December, –20.4°C in January, –19.1°C in February, –13.2°C in March, and –4.5°C in April. The average wind velocity in these months is 4.9 m/sec.

TABLE 5.1
Defining the surface modulus of basic structural types of different shapes

Structure	Surface modulus
Columns and rectangular beams with sides b_1 and b_2 (m)	$M = \frac{2}{b_1} + \frac{2}{b_2}$
Columns and box beams with side b	$M = \frac{4}{b}$
Cube-shaped structures (e.g., foundation) with side b	$M = \frac{6}{b}$
Structure shaped as parallelepiped with sides a, b, and c	$M = \frac{2}{a} + \frac{2}{b} + \frac{2}{c}$
Structure shaped as parallelepiped and adjacent to massive structure	$M = \frac{2}{a} + \frac{2}{b} + \frac{1}{c}$
Slab or wall of thickness a	$M = \frac{2}{a}$
Structure shaped as solid cylinder of diameter d and height h	$M = \frac{4}{d} + \frac{2}{h}$

An appropriate coefficient of external heat exchange or its inverse value — thermal resistance to heat exchange at the interface of the formwork and the ambient air — is used to allow for the effects of wind.

The insulation cover of the formwork and of the exposed part of the structure should be windpoof and protected against wetting that may be caused by snowfalls or rains (during thaws).

Concretes for thermos cured structures. It is recommended that concretes for massive structures with M > 3 be mixed with highly alite and rapid-hardening Portland cements. Low-heat cements, including Portland blast-furnace cement, Portland pozzolana cement, and Portland belite cement, generally are used to concrete structures with M below three.

Accelerators, such as calcium chloride $CaCl_2$, calcium nitrite $Ca(NO_3)_2$, or sodium nitrite $NaNO_2$ combined with calcium chloride, may be added to make concrete harden faster at a low ambient temperature. All concretes for massive structures are mixed with a plasticizer to minimize the water content while producing a mixture of the required plasticity.

The temperature of concrete to be placed in a massive structure should be as high as possible. However, when concrete is delivered from a central batching plant, its temperature in winter as discharged from a mixer must not be higher than 30 to 35°C to avoid rapid thickening during transportation and the resultant problems with placing and compaction. If concrete is mixed at the site and can be placed within 15 minutes after it was discharged, its temperature can be 60 to 80°C. Since it is difficult to obtain concrete heated to such a temperature, the best practice is to have a colder mixture (10 to 15°C) to be heated electrically at the concreting site just before it is placed into a form.

Preparing the base for a structure to be concreted and placing of concrete. It is very important to prepare a proper base for placed concrete when the thermos method is used. The concrete cannot be allowed to freeze at the interface with the base and no deformation of the base can be allowed when the base is bare ground and frozen at that.

Remove ice, snow, or oil spots from the base. When the base is old concrete, remove cement film, the formwork or its parts, and clean the surface of dust and debris to ensure reliable adhesion of the next concrete lift. It is desirable that the base surface be blown over with compressed air before the concrete is placed.

Frozen concrete or rock (freezing only in winter) is best heated to an above-zero temperature to a depth of 300 mm. If the rock base is in the permafrost layer, it should be warmed to 500 mm. Heaving soils also should be heated to the that depth. There no need to heat the base if the concrete contains an antifreeze admixture.

The base is heated in enclosures made of tarpaulin, plastic film, or any other material. Air heaters, infrared heaters, or any other heating devices are installed inside. Electrodes can be used to heat the ground base with a layer of sawdust wetted with electrolyte between them because frozen ground does not conduct electricity. The sawdust heats and causes the upper layer of the ground to thaw. It become conductive and the ground is warmed up further rather rapidly. The ground base must not be heated by spraying hot water or using a chloride solution.

When the concrete base if heated, its surface temperature should not exceed 90 to 100°C to avoid deterioration of the concrete.

Costs can be saved in some cases by protecting the base from freezing if concreting is planned early in winter. Cover the base with an insulation material protected against wetting and having an appropriate thermal resistance designed for the expected frost. The insulation should be placed so as to overlap the areas of the base adjacent to the concrete to be heated by at least 1 m on all sides.

Where the temperature of the base is not lower than –15°C and it is to carry a large plain concrete or reinforced concrete massive structure, the heating may not be necessary. The decision to give it up should be preceded by a thermal analysis. The temperature of concrete when a structure is be erected on an unheated concrete or rock base can be found from the nomogram plotted by V.N. Lemekhov (Figure 5.1). The following specifications should be met:

- The thickness of the first concrete lift should be at least 20 cm for the base temperature of no lower than –10°C.
- Each next lift should cover the previous one as fast as possible but not later than within one hour.
- When placed on an unheated base with a temperature lower than –10°C, the temperature of the concrete for ribs and junctions of the unit being erected should be 1.5 or two times as high as the one found from the nomogram.
- When the ambient air temperature is lower than –10°C, the concreting should be done in an enclosure where the temperature should not drop below –10°C.

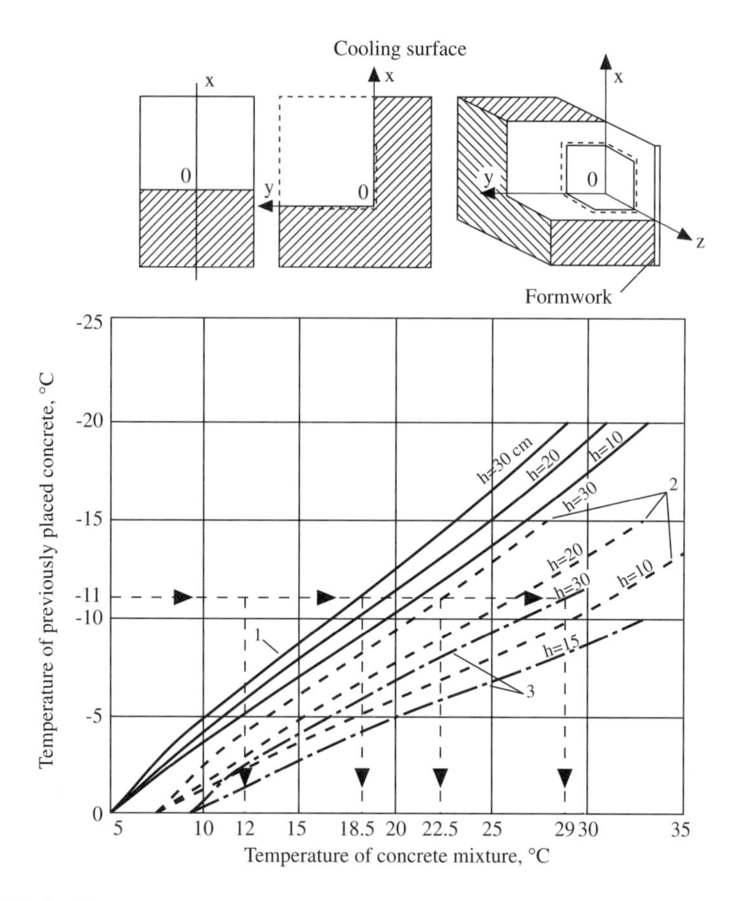

FIGURE 5.1 The nomogram used to determine the temperature of concrete mixture placed on cold base. 1 — one cooling surface, 2 — two cooling surfaces, 3 — three cooling surfaces.

5.2 PLACEMENT AND CURING OF CONCRETE IN STRUCTURES BUILT ON PERMAFROST GROUND

Buildings and structures can be erected on permafrost in two ways. The base ground can remain either frozen or thawed during the whole period of service. This should be borne in mind in preparing the base for a reinforced concrete structure.

When the base ground is to remain frozen during the service life of the structure to be erected, the temperature of the base has to be lowered by freezing before concrete is placed. This does not include rock bases in permafrost layers.

When the base ground is to be used in thawed condition, it should be heated before concrete is placed. The base may not be heated when the layer is thin. It will thaw in the process of service of the structure.

Construction on permafrost is governed by a design that indicates also the nature of the ground in the base. The design of a cast-in-place foundation must take into account the thermal interaction of hardening concrete with permafrost. Failure to consider this aspect may have dire consequences, even collapse of or serious damage to the structure because of changes in the permafrost ground condition of the base.

Foundations cast on the permafrost that should remain frozen during their service life need certain time and conditions for the concrete to freeze solid with the ground. The thickness of the thawing permafrost interlayer and the time during which the concrete is to freeze solid with the ground can be found by calculations, sometimes quite complex ones, that are not described in this book. They are given in detail in regulations that govern construction on permafrost.

It should be borne in mind in concreting that concrete continues to harden slowly at the interface with the frozen ground that has a temperature below –3°C. So the foundation can be loaded only after the frozen condition of the ground, disturbed by the installation of the foundation, is restored, and the concrete should reach its design strength by that time.

The permafrost concreting procedures depend to a great extent on the foundation depth that can be divided into three areas, including:

- an active layer 0 to 3 m deep whose temperature depends on seasonal variation of the ambient temperature;
- an intermediate layer 3 to 10 m deep where the seasonal variation is not great;
- the layer of stable temperatures more than 10 m deep.

Different requirements are placed on concrete according to the layer where it is to be placed. No additional demands for density and frost resistance are imposed on concrete in the layer of stable temperatures and in the intermediate layer. The concrete designed for service in the active layer should meet more stringent requirements for density and frost resistance that should be at least 300 cycles.

Concretes for structures to be erected on permafrost should be mixed with Portland cements. Portland blast-furnace cements and Portland pozzolana cements are not recommended.

Curing is a serious problem faced by concreting on permafrost. A structure may be erected in two ways: (1) in a pit without any contact of its sides with permafrost or (2) it can thrust against it. In the former case, a cushion of crushed stone, dry sand, or slag is made between the bottom of the structure (usually foundation) and the ground to serve as a base for concrete. This insulation cushion prevents the permafrost base from thawing when the concrete is cured. Thermos curing is used for concrete in a form (if the structure is massive enough) or the concrete can be heated by any method.

The situation is much more difficult when the structure thrusts against permafrost. The method of expanded thermos combined with a chemical admixture to accelerate hardening is the best for these conditions. It should be remembered, however, that the admixture may migrate into the ground, depress its freezing point, and thus lower the bearing capacity of the foundation. 1 or 2% of chloride by cement weight are used to accelerate hardening. The quantity of chloride may be increased to 3% in the foundations that are not reinforced or have little reinforcement. The migration of salts does not matter for rock ground. An insulation layer is made between the bottom of the foundation and the permafrost base just like in the first case. The ground at the sides may thaw to a certain depth due to heat transferred from the concrete and migration of moisture. However, as the cement hydration is completed and the concrete cools down until it freezes in the hardened state, the ground gradually returns to its frozen condition. This restoration of the degraded ground may take from several days to several months depending on the temperature of permafrost and the thickness of the thawed layer.

As for foundations of a small cross-section in permafrost, e.g., pile foundations, they are better be made of precast reinforced-concrete components to be lowered into drilled holes. The gap between the ground and a precast unit is filled with liquid mud whose temperature is close to 0°C. When the temperature of the permafrost is –5°C, the freezing may last two or three months or it may take six to 12 days if the temperature is –3°C.

Concreting in permafrost, just like any construction in these conditions, is a very difficult engineering task. We may say that its problems have been solved successfully by now but they need a more thorough and detailed description in a work specifically devoted to the subject.

5.3 CURING AND COOLING OF CONCRETE IN MASSIVE STRUCTURES

The thermos curing of placed concrete produces a nonuniform temperature field over the cross section. So stresses arise in the structure because the rate of concrete hardening is different in different places. But the thermos curing also creates a favorable state of thermal stress that usually prevents cracking at the surface.

A very interesting picture emerges when we analyze the temperature field in a cast-in-place massive structure. The heat starts to drain away from the outer layers to the cold formwork and the insulation material after concrete was placed, compacted, and its exposed surfaces covered. So the temperature in the outer layers drops and the concrete hardens slower. Heat evolution makes itself felt in the interior layers after a few hours and the temperature of the concrete rises. The core warms up gradually by the evolved heat, the temperature of a massive structure with a low surface modulus may rise from 10 to 60°C, and then the structure cools down very slowly. Remember that the hydration heat evolves along an increasing curve (Figure 5.2) and then goes gradually downward. The hydration of 1 kg of cement produces 80 kcal of heat.

The core of the structure expands as it grows warm and so tensile stresses arise in the surface layers of the concrete. But the concrete at the surface still is at an early stage of hardening and thus has a sufficient plasticity to be subjected to strains without cracking. It hardens gradually in this state.

As the temperature in the concrete core decreases, the latter gradually contracts according to the coefficient of cubic expansion, and compressive stresses arise in the surface layers of the concrete. The concrete at the surface resists the stresses

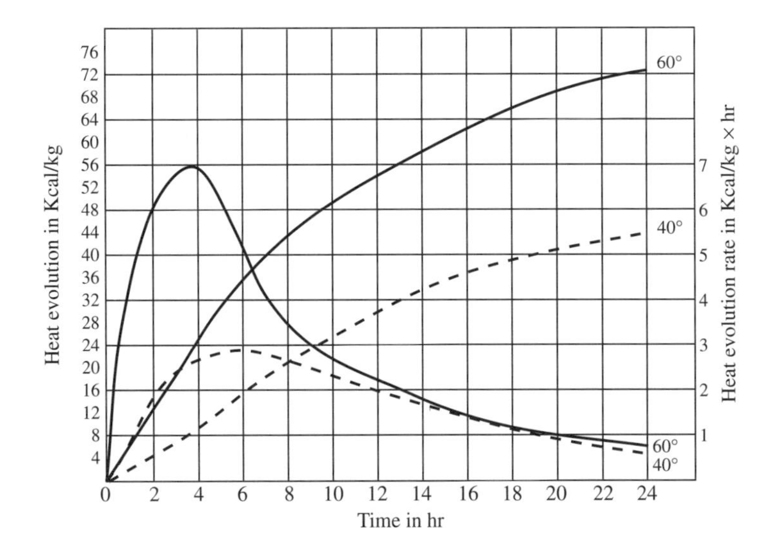

FIGURE 5.2 The integral and differential heat evolution of Portland cement at different temperatures.

without any damage to its structure. On the contrary, it becomes denser and no cracks can be seen in it.

When it heats and cools down, the concrete hardens in the massive structure at different temperatures. The strength development rate is shown in Figures 5.3 and 5.4. Since it is uneven, the average temperature is established from individual cooling intervals. The length of an interval is assumed to be equal to the time during which the temperature of concrete changes by not more than 10°C. The strength of concrete in a structure is found from measurements of the temperature of hardening concrete. The curves of strength development in Figures 5.3 and 5.4 can be used when the concrete volumes are not great. We can determine the concrete strength more accurately in different parts of a cast-in-place structure from the isotherms of strength development by calculating the average temperature of the concrete for a time interval where the temperature drop does not exceed 10°C. The strength is counted on the Y-axis in % R_{28} along the appropriate temperature curve for the given interval. The transfer to the next average temperatures of the hardening is done in parallel to the X-axis. The time is counted by summing its intervals corresponding to average temperatures (Figure 5.5).

Insulation and form removal. The thermos method usually requires an insulation of the formwork to ensure the required strength in the surface layers of a structure by the time when the forms are removed. The thermal resistance of the forms for structures with a surface modulus lower than two is found for end flat and rounded faces and for sides. When the end faces are covered, the insulation should overlap the sides approximately by two m. The required thermal resistance of windproof insulation for end and side faces according to the ambient temperature can be found from a curve (Figure 5.6) as applied to concrete with a temperature

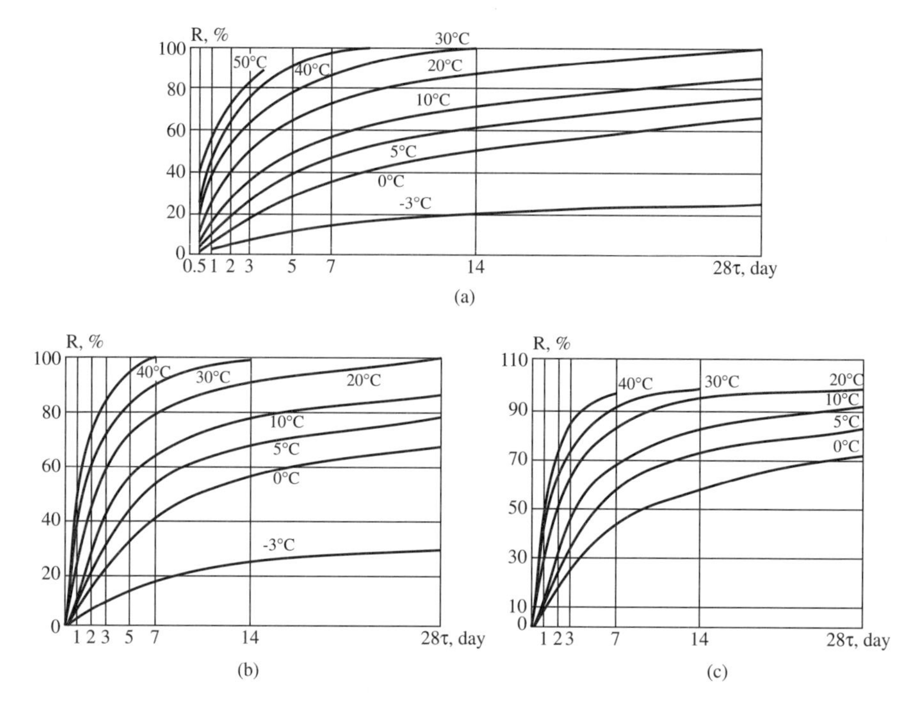

FIGURE 5.3 Strength development of Portland cement concrete (in % R28) at temperatures from –3 to +40°C. a — concrete of grade 200–300 with Portland cement of grade 400, b — concrete of grade 400 with Portland cement of grade 500, c — concrete of grade 500 with Portland cement of grade 600.

higher than 15°C. When the temperature (t) of the concrete being placed is lower than 15°C, the thermal resistance values for the insulation of ends and sides of the structure is multiplied by $\frac{15}{t}$ or is found by calculations.

If the insulation has a higher thermal resistance than an optimum value, the permissible difference between the temperatures of concrete and of the ambient air should not exceed 27°C. It is dictated by the need to make the concrete structure crack resistant.

Concrete is placed as fast as possible without intervals unless the design does not require to break the structure into separate units. The temperature variation in concrete lifts should not exceed ±5%. When the concreting is completed, the top of the structure is covered with an insulation having an appropriate thermal resistance. In exceptional cases, when the concreting process was interrupted for some reason, the difference between the temperature of placed concrete and that of the one to be placed should not be higher than 15°C.

For structures with a surface modulus higher than 2, which should meet particularly stringent requirements for crack resistance, the formwork can be removed at the temperature difference between the exposed surfaces of the concrete and the air found from the formula

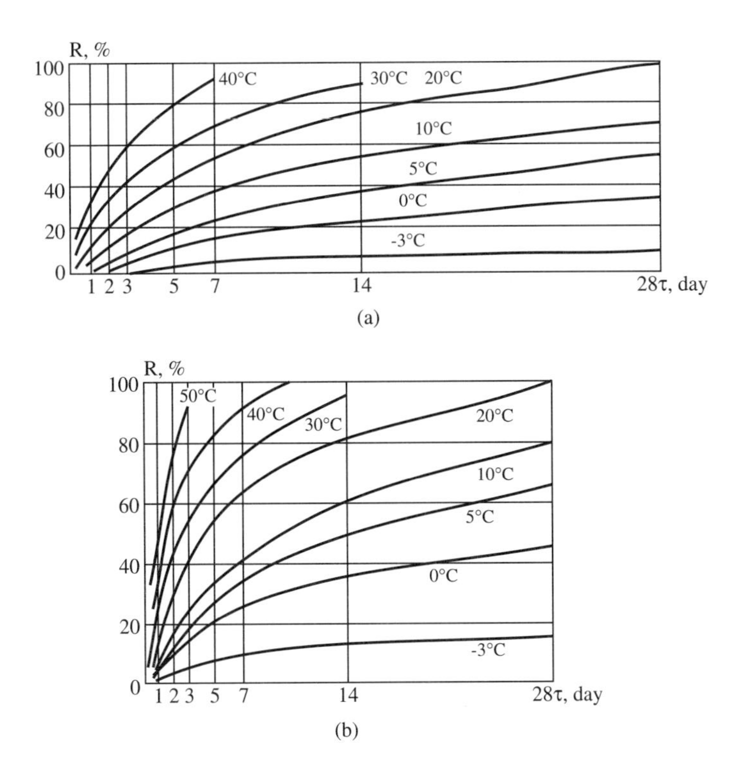

FIGURE 5.4 Strength development of concrete with Portland blast-furnace cement (in % R28) at temperatures from –3 to +40°C. a — concrete of grade 200 with Portland blast-furnace cement of grade 300, b — concrete of grade 200-300 with Portland blast-furnace cement of grade 400.

$$\Delta t = \varepsilon\left(128 + \frac{\beta M}{3 + 10\sqrt{V_{max}}}\right)$$

where: ϵ = permissible tensile strain of concrete by the time of form removal; it can be found either from laboratory experimental data or assumed to be 0.11 mm/m for normal weight concrete and 0.15 mm/m for lightweight concrete,

β = shape factor. It is 132 for structures with ribs, such as joists or T-beams, and 380 for structures without ribs, such as cylinders or shells,

V_{max} = maximum wind velocity in m/s as forecast by a weather station,

M = surface modulus in m^{-1}.

For structures whose surfaces are protected by a detachable insulation, the permissible temperature difference between the concrete surface and the air is given by

$$\Delta t = \varepsilon\left[128 + \beta M\left(1.16R + \frac{1}{3 + 10\sqrt{V_{max}}}\right)\right]$$

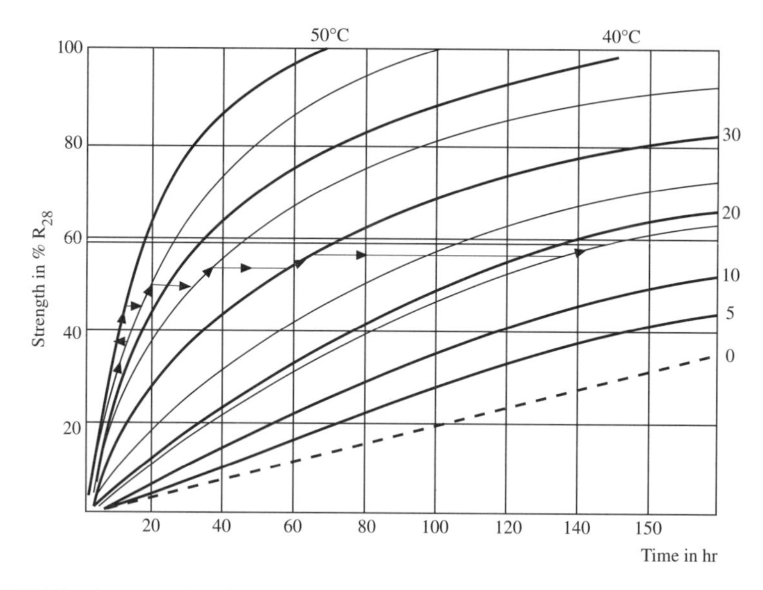

FIGURE 5.5 An example of evaluating the strength of concrete that hardened at different temperatures.

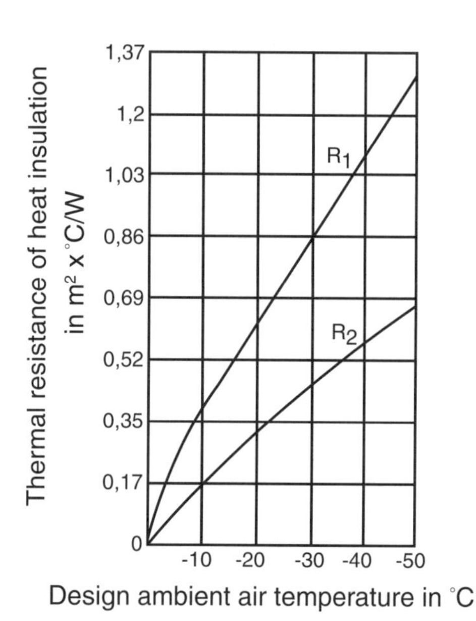

FIGURE 5.6 Determination of optimum thermal resistance of heat insulation of end faces (R_1) and sides (R_2).

where: R is thermal resistance of the insulation in $m^2 \times °C/W$.

Studies have shown that for Portland cement concrete of grade 400 and the rise of the initial temperature from 25 to 50°C the thermal resistance of the insulated formwork can be reduced two or 2.5 times, and the time of curing the concrete in the form can be shortened by 20 to 35%. The initial temperature is usually increased to 50°C or more by preheating the concrete mixture before placing. A shorter curing time in fact makes it possible to increase the turnover of forms. And the use of concrete preheated to a high initial temperature allows in some cases to do without the insulation.

The form removal time for structures where concrete is cured by the thermos method depends on the requirements for the strength of the concrete as with any other curing method. The only thing that should be taken care of and borne in mind when the forms are removed is the temperature difference between the surface layers of the concrete and the air because the problem of cracking is particularly important for massive structures.

Determining the concrete cooling time. An analysis of the cooling of concrete in a massive structure in the frost is a very complex task. The nonuniformity of the temperature field causes the concrete to harden unevenly in different parts of the structure. In addition, many other factors affect the cooling rate, including:

- the temperature of the concrete when it is placed (initial temperature);
- ambient temperature;
- the type of cement used and its heat evolution rate and content in 1 cu.m of concrete;
- the type of formwork;
- insulation and its thermal resistance;
- thermal physical properties of plain or reinforced concrete;
- the size and shape of the structure;
- the velocity and angle of attack of the wind.

There may be other influential factors (the presence of chemical admixtures, the presence of many protruding reinforcement bars, etc.) that should be also allowed for in the analysis.

The first consideration in calculating the cooling of placed concrete is provision of cooling conditions such that the concrete develops the required strength by the time of form removal rather than recording the kinetics of the cooling. Then the analysis of the thermos curing of the concrete includes:

- the cooling time of concrete and the strength it should attain by the time it freezes;
- an optimum thermal resistance of the insulation to produce a favorable state of thermal stress in the structure and to maintain the temperature required for the concrete to develop the necessary strength by the freezing time;
- the time of form removal.

There are several methods used to calculate the thermos curing of concrete in a structure. All of them are fairly approximate because the cooling of placed concrete is a three-dimensional problem of thermal physics, which is hard to solve. The easiest way is to determine the time of cooling and development of the required strength of concrete from curves or nomograms plotted for certain types of structures and specific curing conditions. If these are not available, the analysis can be made by a computer provided with necessary data, including forecasts (e.g., variation of temperature and of the wind during curing). An elementary calculation of the concrete cooling time can be made by using a formula proposed by Prof. B. G. Skramtaev and refined by Prof. S. A. Mironov. This method estimates the temperature on the outer surface of the formwork taking into account heat radiation losses. The cooling time is given by

$$t = \frac{c\gamma(t_1 - t_2) + q\text{Ц}}{KM(t - t_3)}$$

where: t = concrete cooling time in hrs,
c = specific thermal capacity of concrete in kJ/kg × °C,
γ = volumetric weight of concrete in kg/m^3
t_1 = initial temperature of concrete before placing in 0°C,
t_2 = end temperature of concrete to which the cooling is calculated in 0°C,
q = heat evolution of 1 kg of cement during the cooling of concrete (Table 5.2) in kJ,
Ц = cement content in 1 m^3 of concrete in kg,
K = heat transfer coefficient of the formwork in W/m2 × °C,
t = average temperature of concrete during cooling in °C,
t_3 = ambient temperature in °C.

This formula can determine the duration of cooling of concrete in a structure and its strength by the time the concrete reaches the temperature on the surface of the structure, to which the cooling time is calculated. To use it, we should know the size of the structure, steel content in 1 cu.m of the concrete, the type and grade of the concrete, the type and reactivity of cement and its content in 1 cu.m of the concrete, the ambient temperature, wind velocity, the design of the formwork, and the initial temperature of the concrete.

The cooling analysis proceeds as follows: we find the volume of the concrete, then the surface exposed to cooling and the surface modulus of the structure, and finally the heat required to warm up the reinforcement and formwork. The results serve a basis for calculating more accurately the initial temperature of the concrete taking into account the heat expended for the reinforcement and formwork. After we calculated the average temperature of the concrete in the process of cooling we use the above formula to find the cooling time of the structure disregarding the heat evolution of the cement. The heat evolved by the cement can be found on the basis of the structure's cooling time and the average hardening temperature of the concrete (Table 5.2 and Figure 5.2). Knowing the cement content in 1 cu.m of the concrete

TABLE 5.2
The heat evolution of cements, q, according to the temperature and time of hardening

Cement type and grade	Temperature in °C	Cement heat evolution in kJ/kg for hardening time in days							
		0.25	0.5	1	2	3	7	14	28
Portland cement 300	5	—	—	25	59	84	168	210	230
	10	8	25	42	84	126	189	231	272
	20	25	42	75	126	168	230	251	293
	40	50	84	147	189	230	251	293	—
	60	84	147	189	230	272	298	—	—
Portland cement 400	5	—	—	29	63	109	188	210	251
	10	13	25	50	105	147	210	251	293
	20	42	67	105	168	210	272	314	335
	40	84	134	189	230	272	314	335	—
	60	130	188	230	272	314	335	—	—
Portland cement 500	5	13	25	42	89	126	189	230	272
	10	25	42	63	105	168	251	293	314
	20	42	84	126	189	251	293	335	377
	40	105	168	210	272	293	356	377	—
	60	189	230	272	314	356	377	—	—
Rapid-hardening Portland cement 600	5	25	34	50	105	147	210	251	314
	10	34	50	75	126	168	372	335	377
	20	63	105	147	210	293	335	377	419
	40	117	189	230	293	335	377	419	—
	60	210	251	293	335	377	419	—	—
Portland blast-furnace cement and Portland pozzolana cement 300	5	—	13	25	42	63	126	168	189
	10	—	25	34	63	105	168	210	230
	20	—	34	63	126	147	210	251	272
	40	42	75	117	168	210	251	272	—
	60	63	105	147	210	230	272	—	—

we can easily calculate the total heat evolved during the hydration of the cement in the period of cooling. The same formula can be used to determine more accurately the structure's cooling time taking into account the heat evolution. Then we can determine the strength that the concrete reached by the time of cooling from the concrete strength development curves.

Where the strength of the concrete by the time it cools down proves to be lower than required, the insulation cover of the formwork must be changed to something warmer with another coefficient of heat transfer.

The effects of protruding reinforcement on the cooling of concrete. When structures are cast in place, they nearly always have protruding reinforcing bars because the concreting, with rare exceptions, is carried out in lifts rather than through the entire height at once. The effects of the protruding reinforcement on the temperature field in placed concrete is of practical importance and appears worth of investigation.

The building code of Russia recommends that reinforcing bars of more than 25 mm in diameter, rigid rolled sections, or large steel inserts be heated to an above-zero temperature before concrete is placed during the erection of cast-in-place reinforced concrete structures when the ambient temperature drops below –10°C.

Studies found that the reinforcement inside massive concrete structures does not affect its cooling to any great extent. We can say that a massive reinforced concrete structure cools down in the same way as the structure made of plain concrete. The same result is produced in studying the impact of the reinforcement on the shape of the isotherm. It changes so little when it crosses the reinforcement that it may be neglected in practice.

The picture is different if reinforcing bars protrude from the concrete surface. Heat losses through them can be quite high and found from the equation:

$$Q = kFt = k\pi dnlt$$

where: F = the area of the reinforcement surface exposed to cooling,
k = coefficient of total transfer of the insulation of the reinforcement; when there is no insulation, $k = \alpha$,
d = diameter of the reinforcing bars,
n = number of bars,
l = distance from the concrete surface to the point of a bar, where the temperature is 0°C,
t = temperature of the concrete.

Studies of the effects of protruding reinforcement on the temperature field in concrete were conducted at the NIIZhB at an ambient temperature from –22 to –34°C. The experiments varied the diameter of the bars, the length of their protruding sections, and the types of cover for the exposed concrete surface. The diameters were 10, 20, 30, and 40 mm, the protruding lengths were 15 and 90 cm. The concrete surface was covered with two layers of bituminous felt and a layer of sawdust, four cm thick in all, or with two layers of bituminous felt and foam plastic five cm thick. The temperature was measured with thremocouples installed the surface of the bars and in the concrete and fastened with nylon-6 threads.

The effects of the protruding reinforcing bars on the freezing depth of water in the areas adjacent to the reinforcement and on the formation of zero isotherms were studied on models where concrete was replaced with water.

An analysis of the distribution of temperature in a concrete specimen covered with bituminous felt and sawdust and having a protruding bar 90 cm long proved that after the concrete was placed in a form with a cold reinforcement the temperature at the top of the specimen where it contacted the bar dropped below zero within 30 min. The concrete froze to a depth of four cm. In two hours after placing the temperature of the concrete at the top of the specimen rose to 0°C due to the high heat conductivity of steel. The temperature at the interface with the bar increased but did not rise above zero. The temperature near the reinforcement dropped after four hours and the freezing went deeper reaching 30 cm in 11 hours. Observations for 21 and 31 hours found that the frozen concrete layer near the reinforcement gradually widened. The subzero temperature continued to drop at the interface with the bar and affected considerably the formation of the temperature field in general and the structure's cooling rate. The temperature dropped near the reinforcement because, first, the reinforcement was cold when the concrete was placed and, second, the heat conductivity of steel is 30 times as high as that of hardening concrete. The high conductivity of steel helps to remove heat from concrete through its protruding part, which depresses the temperature in the layers near the reinforcement and then further in adjacent concrete layers.

The analysis of the distribution of temperature in the concrete covered with foam plastic revealed that after the concrete was placed in a form with cold reinforcement the temperature of the concrete decreased somewhat at the top of the specimen near the reinforcement. The temperature drop was found to a depth of 16 cm at the interface of the concrete with steel. The temperature in the top part of the specimen rose to 7°C and was 4.8 at the interface in two hours after placing. The reason for this slow change in temperature was the cover of the specimen's top surface and partial freezing of the concrete at the interface area. The freezing depth of the concrete near the reinforcement reached 22 cm after 21 hours and increased to 30 cm after 31 hours.

Investigations of the formation of the temperature field in the concrete covered with bituminous felt and sawdust and having a protruding bar 15 cm long found that the concrete started to freeze near the bar at the top of the specimen 21 hours after placing and the freezing depth reached 30 cm by the 31st hour. It means that a short protruding reinforcement slows down the freezing of concrete next to steel but the freezing depth was the same by the end of the experiment as it was when the bar protruded to a length of 90 cm.

Studies that used a model to find the freezing depth of water near reinforcing steel are of interest. Although the diameters of the bars were different, the isotherms had the same conical shape. The freezing depth of water near the reinforcement depends to a great extent on the diameter of the bars. The protruding bar length being 90 cm in all cases, the freezing depth of the water was 6 cm when the diameter of the bars was 10 mm, 12 cm when the diameter was 20, 18 cm for the diameter of 30 mm, and 24 cm for the diameter of 40 mm. So the freezing depth is directly proportional to the diameter of a reinforcing bar.

The above-described experimental data established the functional relationship

$$H = md$$

where: H = water freezing depth at the interface with the bars in cm,
m = constant value equal to 6 (proportionality factor),
d = diameter of a bar in cm.

This relationship can be also used to determine the freezing depth of concrete at the interface with the reinforcement although the factor m will be a little smaller. It depends on the temperature of the concrete, type of cement, etc.

The studies conducted to investigate the effects of protruding reinforcing bars and of their diameter on the formation of the temperature field near steel lead to following conclusions.

The protruding reinforcement greatly affects the variation of the concrete temperature at the interface with the bars and the temperature field in concrete in the vicinity of the protrusions in general. The drop of temperature and the concrete freezing depth depend to a great extent on the diameter of the bars and on the protruding length. When the protrusions are short and the exposed concrete surface is well covered, the rate of the temperature drop decreases. However, when the diameters of reinforcing bars exceed 30 mm, the temperature drops below zero at the interface of concrete with steel. The subzero temperatures at the start of cooling cause thin ice films and ice segregations to form at the interface of the protrusions with the concrete, which loosens the concrete at the interface.

Early freezing of concrete near protruding rebars may decrease its bond with steel, adversely affect the behavior of the concrete structure under dynamic and vibration actions, and facilitate corrosion in an aggressive environment.

It should be noted in conclusion that, regardless of the parameters of curing by the thermos method found by calculations, the temperature of the concrete after placing should be monitored at all times, particularly in the spots prone to fast cooling. It will help to take timely measures to maintain the required above-zero temperature in concrete to let it attain the strength established in the design by the time of form removal.

6 Electric Contact Heating of Concrete

6.1 THE PRINCIPLE AND APPLICATION OF THE METHOD

Electric contact heating of concrete is a common method of thermal treatment used to accelerate the hardening of concrete in cast-in-place structures in summer as well as in winter. Heat is supplied from an electric heater to the surface of the concrete in the structure to be heated by radiation or conduction, where it spreads down into the core of the material owing to its heat conductivity. The heat transfer in electric contact heating does not differ in fact from that in steaming or air heating. So all the laws concerning heating conditions and the formation of the structure of the material remain the same. There are, however, some specific features due to the design of heaters, the manner heat is supplied to concrete, mass exchange, etc.

Electric contact heating of concrete can be used to erect any structure, particularly those whose surface modulus exceeds six and the surface has a well-developed and complex configuration regardless of the reinforcement. A distinction is made in the way heat is supplied to concrete between low-temperature electric heating whereby heat is transferred from a heater to the concrete by conduction, and high-temperature electric heating where heat is delivered by radiation. The temperature on the heating surface of low-temperature heaters is under 250°C. The surface of high-temperature heaters is hotter than 250°C.

Low-temperature electric heating is generally applied by heating formworks and heating blankets (mats), or by heating forms at precast factories. The heating formwork is designed so that an electric heater is placed at the back of the deck and its heat is transferred to the adjacent concrete. To reduce heat loss to the environment, the outer surface of the formwork is insulated, the heat insulator being protected against wetting or damage with plywood or a thin steel sheet. This design of the heating formwork minimizes heat loss so that most of the heat released by the heater is delivered to the concrete. The heating elements should be installed so that the temperature difference on the formwork surface in contact with concrete does not exceed 5°C.

6.2 HEATERS FOR ELECTRIC CONTACT HEATING

Heaters for low-temperature electric heating of concrete have many designs but the most convenient for the heating formwork are string, mesh, rod, strip, and fabric types. Tubular, coaxial, tube-and-rod, and angle-rod heaters are generally not used at construction sites because they are too heavy. The choice of an electric heater depends on the type, size, and configuration of the structure to be heated, available electric power, and whether a particular heater can be obtained or fabricated at the site. Most electric heaters have to be made on the spot.

String heaters. String heaters are made of metals having a high specific resistance, such as nichrome or others. The diameter of the strings is two or three mm. Since these materials are relatively expensive, steel wire may be used. The strings are usually insulated for easier wiring and better electric and fire safety. The heating wire is a convenient type of heater. String heaters are installed at a distance of three or four cm from the deck, the strings being spaced five to 15 cm. A wider spacing may not produce the required uniform temperature distribution over the deck surface.

The heating wire is a typical string heater. It is called *corde de chauffage* in France. The heating wire is convenient for heating any type of reinforced concrete structures. When properly installed, it can ensure a sufficiently uniform temperature field and create a warm envelope around the structure by raising the temperature in the surface layers where it is embedded. It should be noted, however, that an important disadvantage of this heater is that it can be used only once because it remains in the concrete for good. But if the heating wire is inexpensive, it does not involve much additional cost and is economically feasible. The heating wire may be useful for some reinforced concrete structures in service as an embedded heater. It can thus help to make warm floors, e.g., in animal houses, warm walls or other type of structures that should be heated in service.

The heating wire is a steel core of 1.2 to two mm in diameter with plastic insulation that makes it very reliable in operation protecting against current leaks, ensures proper safety during concreting and later during service if the structure is to be heated with the wire. Prof. V. P. Lysov developed a heating wire with a nonmetallic core, which is very flexible, reliable in operation, and costs about the same as the steel heater.

An advantage of the heating wire is that it can heat concrete inside a structure, which is impossible with the surface heating by any low-temperature or high-temperature heater. Since the heater is embedded in a structure, all its heat is transferred to the concrete without loss to the environment because its environment is the concrete. Naturally, the efficiency of the heater is high.

To determine the parameters of electric heating with a heating wire, we must know the electric output that depends on the voltage supplied, the cross section of the conductor, and on its resistance. The voltage generally does not exceed 120 V but in practice it mostly varies between 40 and 60 V. Given the resistance of the wire and the voltage used, the heating wire is installed and connected in independent sections. For example, when the diameter of the conductor is 1.2 mm and the voltage is 60 V, a section may be 30 m long. For a diameter of 1.4 mm the length may be increased to 40 m. Since the commercial plastic insulator is not heat resistant, its surface temperature usually should not exceed 70 or 80°C. A higher temperature can melt the insulator and the exposed wire may short-circuit the reinforcing steel.

The electric parameters of a heating wire, including specific resistance, power consumption, current strength, and the maximum temperature allowed on the surface without damage to the insulation, are determined by the on-job laboratory and are used in calculations to find the electric power needed to heat the structure and the number of heaters required. To produce a uniform temperature field, the heating wires are arranged in parallel strings spaced five to 15 cm.

When the heating wire is used to heat a structure from outside, it is covered with an insulation material. When it is not too cold (down to –10°C), it is enough to cover it with tarpaulin to prevent excessive heat loss to the environment. The heat loss is particularly great in the wind and any cover, even tarpaulin, can reduce it considerably.

The heating wire is installed after the reinforcement of a structure is completed. It is attached to the rebars with common binding wire or adhesion tape spaced as designed so that the heater is not damaged when concrete is placed. The ends to be used as power leads are fixed in a convenient place and covered to avoid damage. After the formwork is installed, the operation of the heater must always be checked first and only then concrete can be placed. After the structure (column, a small foundation, etc.) is concreted, the exposed surface is covered with steam and heat insulator and electric current is passed through the heaters. Where a large structure is to be concreted (a floor slab, the floor in an industrial building, etc.), it is a good idea to divide it into separate areas, each provided with a heater of its own to be energized immediately after placing, compaction, and covering of concrete in the area.

Antifreeze admixtures are better be added to concrete in a severe frost to avoid freezing. Should a superplasticizer be introduced into concrete in mixing, the amount of any antifreeze admixture may be cut by half as compared with the same concrete but without the superplasticizer.

The temperature of concrete, particularly in the spots prone to fast cooling, must be closely monitored in the course of electric contact heating just like with any other method of curing in the frost.

Heating wires were used to a great advantage in casting reinforced concrete structures of Moscow's largest church in winter. Reinforced concrete column bases, columns, floor girders and slabs, and massive pillars were cast of concrete with a slump between 12 and 24 cm (depending on the steel ratio) and containing a superplasticizer and an antifreeze admixture (sodium nitrite). The heating wires were installed in peripheral layers of the column bases, columns, and pillars with a spacing of 15 cm and at the bottom of the floors. The heaters were energized immediately after the concrete was placed and compacted, and the temperature started to rise gradually until it reached 40 to 60°C. After it was heated for 24 hours, the formwork was removed and the structural members were covered at once with insulation panels to retard cooling. In massive pillars the temperature rose to 20 or 40°C (depending on the ambient air temperature) whereupon the heaters were deenergized. The temperature continued to rise due to heat of hydration to 60 or 80°C. The formwork of the pillars was removed after 24 hours and their surface was covered with insulation panels. The latter were removed from all types of structures when the temperature difference between the surface of the concrete and the ambient air dropped to 20°C.

As for the floors, the heaters were energized immediately after concrete was placed and its surface was covered with mineral wool mats placed on plastic film.

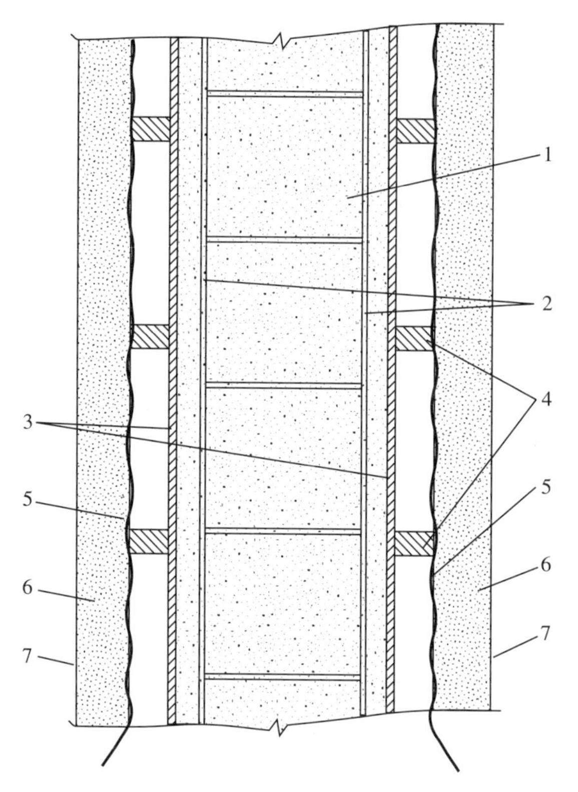

FIGURE 6.1 Heating a reinforced concrete partition in a heating formwork with the deck made of water-resistant plywood. 1 — heated concrete, 2 — reinforcing cage, 3 — deck, 4 — wooden block, 5 — wire heater, 6 — insulation, 7 — protective cover.

The formwork was struck when the strength of the concrete attained 70 to 80% of its design value depending on the span of a beams or a slab.

All the reinforced concrete structural members of the church erected in winter are of high quality and their characteristics are as good as those of similar members erected in summer. Non-destructive methods (three methods) were used for quality control, which ensured a high reliability of the results.

The heating wires have been used successfully for concrete heating in construction of many large buildings and structures in Moscow and became quite popular with builders.

The heating wire is convenient for use in heating formworks where it is installed on the back of the deck. To avoid large heat loss to the environment, the heating formwork is insulated from outside, which makes this heating method cost effective and convenient. Forms with heating wires are widely adopted in Finland, where this method proved its worth as a very efficient technique (Figure 6.1).

The heating wire with a flexible carbon filament is convenient for fabrication of heating blankets used mainly to heat concrete in structures with a large exposed surface not covered with formwork. The blankets placed on fresh concrete previously

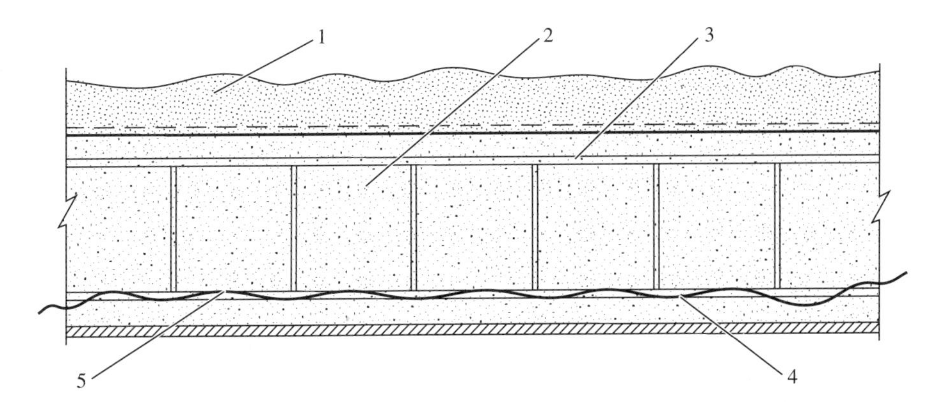

FIGURE 6.2 Combined method used to heat a reinforced concrete floor. 1 — heating blanket, 2 — heated concrete, 3 — reinforcing cage, 4 — heating wire, 5 — formwork.

covered with plastic film can produce very favorable conditions for the hardening of the concrete in the structure.

A combined method is better be used to heat floors and slabs thicker than 20 to 25 cm at low temperatures. The heating wire is installed in the bottom and the concrete is covered with heating blankets on the top (Figure 6.2). The combined method has been tested many times in winter construction and produced excellent results.

Mesh heaters. These heaters are made of metal mesh, e.g., brass, with openings of two or three mm and the wire diameter of 0.35 to 0.5 mm. A mesh may have somewhat different parameters provided the required power is ensured. Mesh heaters made of brass, high-temperature steel, and stainless steel were tested by the NIIZhB. In the power range between 0.2 and 0.5 W/cm^2, current spread over cross sections of mesh heater strips of different widths practically as evenly as in strip heaters made of aluminum foil of the same size. The voltage drop in a brass mesh heater was 1.4 to 2.4 W/m in the operating power range between 0.2 and 0.5 W/cm2. The starting current was 3.2 to 6.6 A/cm at a temperature on the heater of 130 to 260°C. A comparison with mesh heaters of high-temperature steel and of stainless steel proved that the brass mesh heaters had a low specific resistance and so a heater may be quite long depending on admissible voltage. The operating voltage of a mesh heater of the same power made of high-temperature steel was nearly twice as high as that for a brass heater. With the same voltage, the area of the heater made of high-temperature steel was smaller by half than that of the brass heater. The voltage drop in a mesh heater made of stainless steel was 2.5 to 3.7 times higher than in a brass heater. The starting currents in meshes made of these metals were about the same.

For practical purposes, specific resistance is a factor when the material is selected for a heater. For structures with large surfaces to be heated, heaters with a low specific resistance, i.e., the ones made of brass, copper, or stainless steel, are the best choice.

Mesh heaters come in the form of strips 10 to 15 cm wide. Nomograms (Figure 6.3) are used to determine the heating surface area of the formwork. The

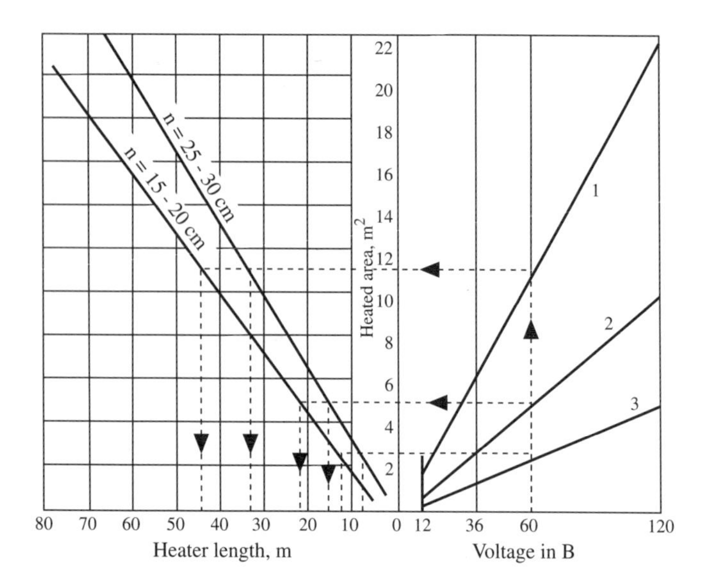

FIGURE 6.3 The nomogram used to determine the heating surface of the formwork according to the spacing of mesh heaters for a mesh width of 15 cm; n is a clear distance between the heaters. 1 — brass, 2 — high-temperature steel, 3 — stainless steel.

nomogram in Figure 6.3 clearly shows that 11 sq.m of surface can be heated at 60 V if we use 34 or 45 m of brass mesh depending on the distance between strips, which is 25 to 30 cm or 15 to 20 cm, respectively. Other conditions being equal, mesh heaters made of high-temperature or stainless steel can heat smaller areas of five and 2.5 sq.m, respectively.

The width of mesh heaters is best to be taken to be 10 to 15 cm in designing a heating formwork. The distance between the strips should be 10 to 20 cm. It can be reduced to two or five cm in the corners of a structure or in other parts that tend to cool down quickly.

Since the mesh heaters are relative newcomers in construction practice and builders are not aware how they are calculated, we describe below basic principles of the design of a heating formwork with embedded heaters of this type.

We determine heat exchange between the heating formwork and concrete taking into account the temperature at the interface, which may have different values from 45 to 85°C and is selected according to the strength of concrete required by the end of heating and other factors. When the heating formwork is designed, the size and installed power of mesh heaters are set in view of electric parameters of the power source.

Just like in any thermal analysis, we determine the amount of heat required for thermal treatment of concrete allowing for heat loss to the environment and for heating of the formwork.

The mean temperature of concrete in heating is given by

$$t_1 = \frac{t_1' + t_2'}{2}$$

where: t_1' = the temperature of concrete at the interface with the heating formwork in °C,
t_2' = the temperature of concrete on the unheated side in °C

We determine then basic parameters of the thermal treatment of concrete in the heating formwork when heat is supplied only from one side.

The specific power required to raise the temperature of one sq.m of the surface of concrete and formwork is found from

$$P = P_1 + P_2 + P_3 = \frac{C_1 g_1 F_1 \delta_1 (t_1 - t_2)}{t_1} + Fk(t_1 - t_3) \quad W$$

where: P_1 = electric power required to heat concrete and the formwork in W,
P_2 = power expended to compensate for heat loss to the environment as concrete is heated in W,
P_3 = power equivalent to the heat released in concrete by cement heat evolution in W, to be included only for structures with a surface modulus Ms ≤ 6
C_1 = reduced specific heat capacity of concrete and formwork material in J/kg × K,
γ = reduced weight of concrete and formwork material in kg/m^3,
t_1 = mean temperature of concrete by the end of its rise in °C,
t_2 = initial temperature of concrete before heating in °C,
t_3 = ambient air temperature in °C,
δ_1 = total thickness of the material of concrete and formwork in m,
F_1 = total area of heating surface in m^2,
F = adopted heated surface of concrete equal to one m^2,
K = heat transfer coefficient of the formwork in W/m^2 × °C,
τ_1 = thermal treatment time in sec

The specific surface power of a mesh heater of an adopted width is found from

$$P' = \frac{P}{10^{-4} KF} \quad W/m^3$$

where: K = coefficient of filling the formwork with the heater found from Table 6.1

Let us find the required length of the mesh heater of the given width as follows:

$$l = \frac{KF}{b} \quad m$$

where: b = width of the mesh heater in m

TABLE 6.1
Coefficient of filling the formwork with heaters K

Distance between heater strips, cm	Width of heater strip in cm				
	10	15	20	25	30
5	0.60	0.72	0.77	0.80	0.85
10	0.50	0.60	0.67	0.71	0.75
15	0.40	0.52	0.60	0.65	0.69
20	0.33	0.40	0.50	0.55	0.60
25	0.32	0.41	0.48	0.55	0.60
30	0.29	0.39	0.44	0.50	0.55

It is more convenient to use nomograms (Figure 6.4) if we want to determine approximately drops of the voltage U and of the current strength I in a mesh heater one m long. They were plotted according to the length of the heater strip and specific power required.

The required design voltage in a single-phase circuit is given by

$$U = lU_1 \quad V$$

where: U_1 = voltage drop found in nomograms in V/m

The required active power of the heater is found from

$$P = IU\cos\alpha \quad W$$

where: I = design current strength found in nomograms in A,
$\cos\alpha$ = phase angle in deg or rad

A transformer is selected and the cross sections of leading wires are determined according to the operating current. With the star connection of the heater to the power source

$$I_1 = I_2 = I; \quad U_2 = \frac{P_a}{\sqrt{3}\ I\cos\alpha}$$

With the delta connection of the heater to the power source

$$I_1 = \sqrt{3}\ I_2; U_2 = U_3$$

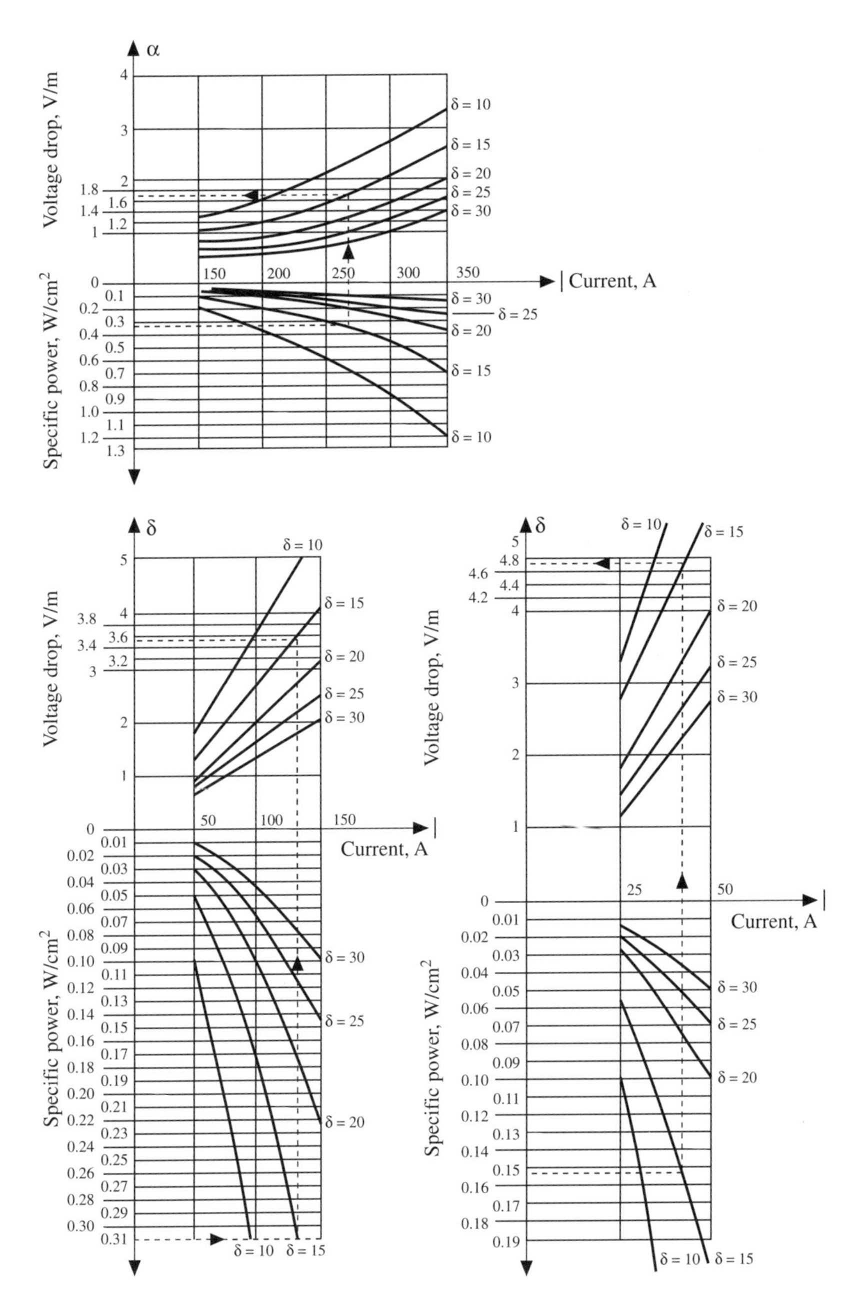

FIGURE 6.4 Nomograms used to find electric parameters of electric mesh heaters according to installed power. A — brass, B — high-temperature steel, C — stainless steel, δ — distance between heaters in cm.

where: U_2 = linear voltage between any two wires of a three-wire three-phase line in V,
I_1 = linear current in one of the wires of a three-wire three-phase line in A,
U_3 = phase voltage between any wires the neutral point of the star in V
I_2 = phase current passing in a side of the delta or in a ray of the star in A.

The design technique may be changed according to the problem to be solved.

Having completed the calculations, which are approximate anyway, we need to check the rate of temperature rise by experiments. Should the experimental data deviate from design values, the latter should be adjusted accordingly.

Special care must be taken to properly connect the strips of mesh heaters as a heating formwork is fabricated. Copper or brass clamps whose specific resistance is lower than the current resistance of the strips are used as connectors. After the mesh strips are laid out and connected they are connected to the power source whose voltage corresponds to the design current strength. Then the actual strength of current is measured. It should correspond to its design value. If there is a difference, the connections are checked and the fabrication of the formwork panels is completed only after the required parameters were provided.

Asbestos sheets no less than one mm thick or any other insulator that can operate reliably at a temperature 1.2 to two times as high as that on the surface of the mesh heater and of the connectors is used to insulate the heater from the deck.

With mesh heaters it is necessary to assign correctly the distance between the heater strips to ensure the formation of a uniform temperature field in the structure to be heated. It is impossible, of course, to have an absolutely uniform distribution of temperature over the cross section of the structure when it is heated electrically and from one side at that. Nevertheless, it is possible to minimize the scattering of the temperature values. For this purpose, we should space the heater strips as close as possible. Ideally, the strips are best be placed without any gap between them but then too many meshes would be needed. Studies found that the distance between strips should be about half of the width of a strip. The latter is generally 10 to 15 cm.

Differences occur in the temperature field of a concreted structure as it is heated. Later on, in the course of isothermic curing, the temperature levels off in the concrete layers that are in parallel to the heater. When a structural member no thicker than 10 or 12 cm is heated from one side, the temperature even on its opposite surface does not differ much from that of the concrete adjacent to the heater. However, the picture is much different when the thickness of a member is 20 cm or more. Although the temperature differences are not great inside the concrete, the temperature on the surface opposite to the heater is two or more times lower. This is the reason why it was established that one-sided heating can be effective only for a member no thicker than 20 cm provided it was insulated from the opposite side. Thicker structures are best be heated on two sides (Figure 6.5). The temperature field in this case is more uniform. Heat evolved as the concrete hardens also is a great help.

This can be seen very clearly in Figures 6.6 and 6.7 that show heating of floor slabs. The concrete was heated in a metal formwork with embedded mesh heaters. The exposed surface of the floor was covered only with a plastic film. The ambient

FIGURE 6.5 Heating concreted floor with mats having electric mesh heaters installed on the concrete surface.

temperature was +10°C because the floor was heated inside a building whose openings were closed and an above-zero temperature was maintained inside. Under a subzero ambient temperature, when the exposed surface is not insulated, the temperature differences increase sharply. So one-sided heating of concrete is not effective without insulation of the opposite surface.

Proper design of the heating formwork with mesh heaters and proper management of the heating can ensure high quality of erected reinforced concrete structures.

Rod heaters. These are made of metals of high resistance or of ordinary steel. The diameter of the rods usually does not exceed five or six mm. The rods form frames that are fixed in the formwork at a distance of five or six cm from the deck. They are well insulated from the frames with insulators made of porcelain or heat-resistant plastic. The rods are spaced eight to 15 cm. This distance may be shorter in the corners and other spots that cool down most of all. The temperature on the formwork surface that touches concrete should be around 80°C in all cases.

Despite the effectiveness of the rod heaters and their long life they consume a lot of metal and are heavier than other types of heaters. For this reason they are used more seldom in cast-in-place structures than string or mesh heaters.

Strip heaters. Heaters in the form of metal strips 0.5 to two mm thick and 10 to 50 mm wide are made of the same materials as the rod heaters. The strips are usually insulated with heat resistant rubber, asbestos fabric, or heat resistant plastic. They may be installed directly on the deck with a spacing that ensures the deck surface temperature of 80°C with temperature differences in its plane of no more than 5°C. The strip heaters are convenient, effective and can form a uniform temperature field at the interface with concrete just like mesh heaters. The design of

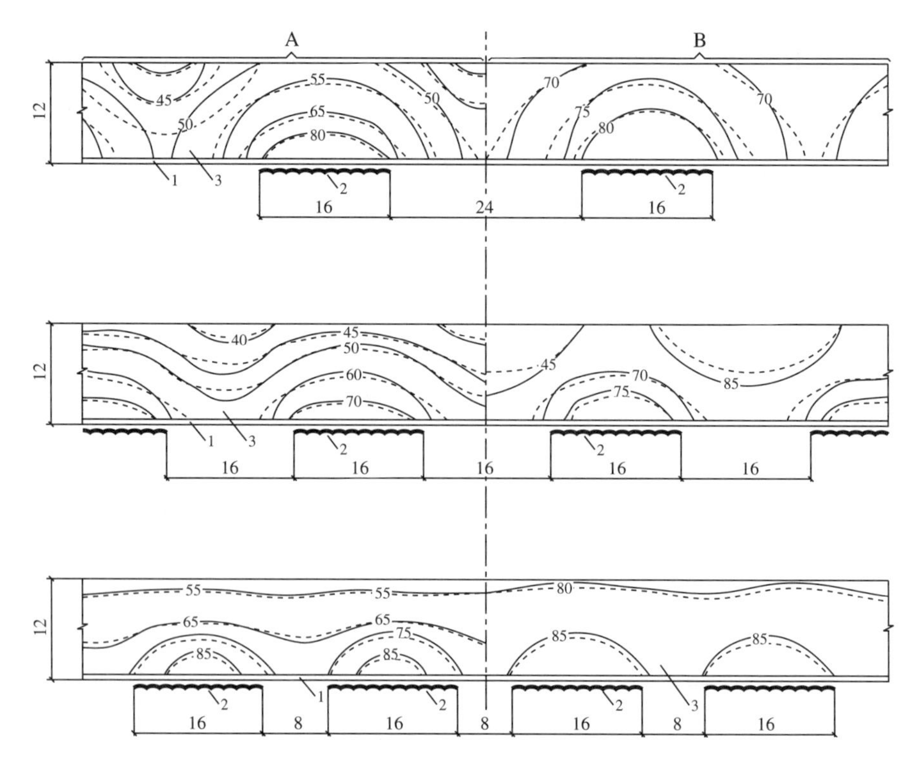

FIGURE 6.6 Temperature fields in a member 12 cm thick heated by electric mesh heaters with different distances between them. A — during temperature rise, B — during isothermal curing. 1 — isotherms, 2 — mesh heater, 3 — concrete.

strip heaters is the same as for mesh heaters but the strips are easier to connect to each other and to the feed wire.

Fabric heaters. Fabric electric heaters are an interesting type. They have the form of strips of carbon fabric, which are installed in the formwork and must be insulated. They are practically no different in action, layout in the formwork, and calculation methods from mesh and strip heaters. The carbon fabric conducts electricity quite well and can withstand heating to 500°C. The most difficult thing with the fabric heaters is their connection with each other and with the feed wire. This problem was resolved by Prof. V. P. Lysov who provided a simple and reliable connection that does not allow contacts to burn.

Using carbon fabric as a basis, Lysov developed excellent electric heaters in the form of heating wires, strips, and plates of heat-resistant plastic up to three mm thick where the heaters are embedded.

Heating blankets. When we erect structures with a large exposed surface in cold weather, there is always a problem, in addition to protecting concrete from freezing, also of heating. Flexible low-temperature electric heaters called "heating blankets" were developed and are used for this purpose. The heaters in the blankets are carbon strips or filaments, thin flexible wire up to 0.5 mm in diameter, a steel mesh embedded in heat-resistant rubber, or other types of flexible heating elements.

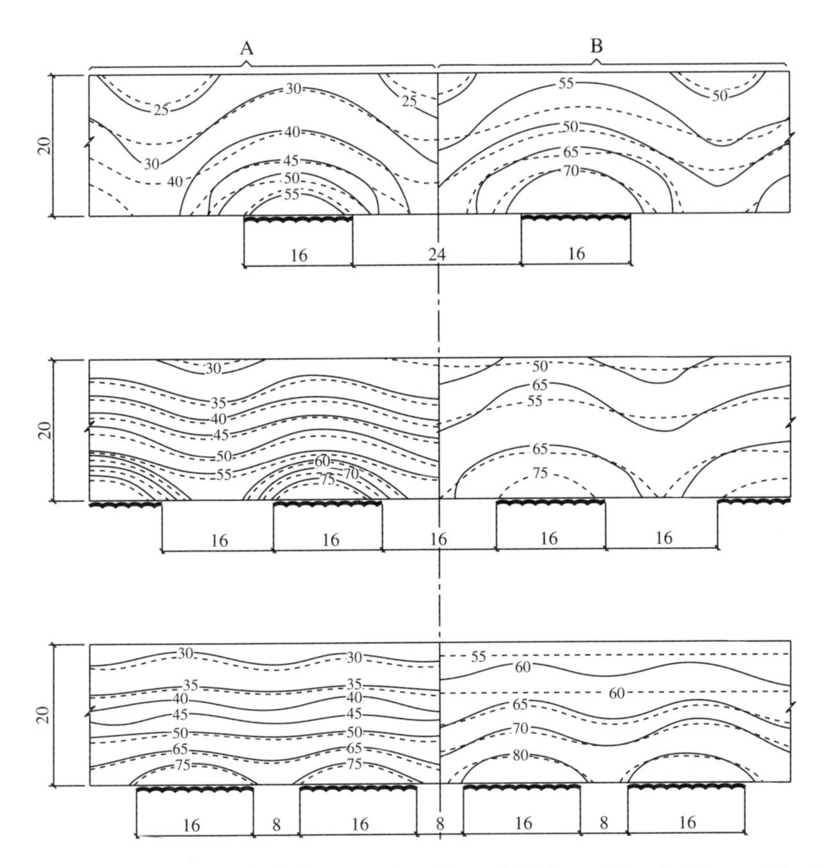

FIGURE 6.7 Temperature fields in a member 20 cm thick heated by electric mesh heaters with different distances between them. A — during temperature rise, B — during isothermal curing. ---- isotherms found by experiments, —— isotherms found by calculations

There are two types of flexible blankets: without or with heat insulators. The first type of heating blankets is a steam-insulated sheath made of heat resistant rubber or plastic that contains inside a heater with electric insulation. These blankets are compact, light, and are easy to handle. Their thickness generally does not exceed several millimeters. To reduce heat loss to the environment in heating, heat insulators in the form of flexible mats are placed on top of a blanket. The mats can be easily rolled out and then rolled up again.

The second type of heating blankets has a steam-insulated sheath too but in addition to heating elements a heat insulator in the form of a layer of mineral wool or another flexible insulation material is made part of it. Such heating blankets are up to 10 cm thick or even more and protect concrete against cooling quite well when placed on top of it and minimize heat loss to the environment in heating. The heating blankets are installed on the surface of placed concrete immediately after evaporation of the surface water film and cover tightly the surface of the concreted structure.

The heating temperature, just like with any other heating method, should be at the level of 60 to 80°C (at the interface with a heating blanket) and the heating should continue until the concrete attains 50 to 75% of its design strength. The blankets are removed when the temperature difference between the environment and the concrete reaches 20°C.

The heating blankets are convenient to use in construction of road and airfield pavements, floors in industrial buildings, sports arenas, and other structures with large exposed surfaces.

The heating formwork provided with flexible electric heaters of conductive carbon fabric ensures a uniform temperature field on the deck surface under any temperature conditions while consuming relatively little energy. The comparatively low temperature of the heater surface (150 to 230°C) makes it much easier to select the type of electric insulation. Fabric heaters increase the weight of the formwork but not much: no more than by 0.4 kg/m2. The heater strip together with the electric insulation is three to four mm thick and the formwork itself has the same dimensions as without heaters.

Infrared heaters. High-temperature electric heaters also called infrared heaters can radiate energy in a wide range of electromagnetic waves from 0.76 to 1000 μm. However, builders use short-wave and middle-wave generators operating in a range of 0.76 to six μm (at a heating temperature of 600 to 2700°C). The radiation density of the heaters rises dramatically at temperatures higher than 250°C and so we can assume that heaters with a temperature of the radiating surface over 250°C are high-temperature or simply infrared heaters.

The latter include carborundum radiators with a heating temperature of 1300 to 1500°C, selite radiators with a heating temperature up to 2000°C, paraboloid radiating lamps with a heating temperature up to 3500°C, and tubular radiators with a heating temperature up to 700°C (coils, coaxial heaters, thermal electric heaters). Infrared rays have a very low penetrability with respect to concrete (less than one mm) and so cannot affect it in any way except heating.

The infrared heaters are used to warm up frozen concrete surfaces, to preheat edges of precast reinforced concrete components to be joined and of steel in the joint area, and to heat placed concrete. To direct radiation, the infrared heaters are used in combination with spherical or trapezoidal reflectors. The former are employed when energy is to be transmitted to a distance of three m and the latter up to one m.

The infrared heaters have been adopted for concrete heating in several countries. They usually come as boxes of different designs and sizes and are installed at a distance from the formwork surface so that the temperature at the interface with concrete does not exceed 80°C. Naturally, a large part of the radiated energy is lost for heating the air, taken away by wind, or is just dissipated.

The calculation of infrared heating and selection of a radiator are based on the required power, the number of radiators, and their layout relative to the surface to be heated. The energy and geometric parameters of infrared units should meet the need to create the required illumination found by calculation. Physical features of infrared radiation must be taken into account in the process. The radiation energy that falls on a body being heated is absorbed and converted to heat only in part. Some energy is reflected by the body surface and some may pass through it. The

TABLE 6.2
Emissive factors of full normal radiation from some materials used in construction

Material	Temperature in °C	Emissive factor
Aluminum:		
polished	50 to 500	0.04 to 0.06
heavily oxidized	50 to 500	0.2 to 0.3
Concrete	20 to 100	0.65 to 0.85
Water layer over 0.1 mm thick	0 to 100	0.95 to 0.98
Wood	20	0.7 to 0.8
Iron:		
polished	400 to 1000	0.14 to 0.38
oxidized	100	0.74
oxidized	125 to 525	0.78 to 0.82
shiny galvanized sheets	30	0.23
oxidized galvanized sheets	20	0.28
Old tin plate	20	0.28
Expanded-clay concrete	20 to 150	0.7 to 0.9
Paints:		
aluminum paints of different ages	100	0.3 to 0.35
oil paints of different colors	100	0.92 to 0.96
Black mat varnish	40 to 100	0.96 to 0.98
Smooth ice	0	0.97
Snow	—	0.8
Steel:		
polished sheets	950 to 1000	0.55 to 0.61
with rough flat surface	50	0.95 to 0.98
heavily oxidized	50 to 500	0.88 to 0.98

body that absorbs all incident radiation and reflects none is called "blackbody." There are no ideal blackbodies on earth and all actual bodies can be called "gray bodies." Blackbodies in space are black holes that absorb everything and radiate nothing thus becoming invisible.

The radiation properties of any gray body are characterized by its radiating power or emissive factor. The latter shows how many times the integral radiation density of a body is lower than that of the blackbody at the same temperature. The emissive factor of the blackbody and its absorbing capacity are equal numerically.

The emissive factors of some building materials are listed in Table 6.2.

The required illumination is found from following formulas for different heating stages:

At the stage when temperature is rising:

$$E_1 = \frac{0.75}{\varepsilon F} P_1$$

At the stage of isothermal heating:

$$E_2 = \frac{1.25}{\varepsilon F} P_3$$

where: E_1 and E_2 = illumination
ϵ = emissive factor of the surfaces that absorb incident infrared radiation (Table 6.2),
F = area of the surfaces that absorb incident infrared radiation in m^2,
P_1 = electric power required to heat concrete in kW/m^2,
P_3 = electric power expended to compensate heat loss to the environment in the process of concrete heating in kW/m^2.

The electric power needed to heat concrete is given by

$$P_1 = \frac{C_1 \gamma_1 V_1 (t_1 - t_3)}{864 \tau_1} + C_3 \gamma_3 \delta_1 F \frac{t_1 - t_3}{2} + (\alpha F_1 + kF_2)\left(\frac{t_3 - t_1}{2} - t_4\right) - ЦE_1 V_2$$

The electric power needed for the period of isothermic heating is given by

$$P_3 = (\alpha F_1 + kF_2)(t_1 - t_4) - ЦE_2 V_1$$

where: C_1 and C_3 = specific heat capacity of concrete and steel, respectively, in Kcal/kg × °C,
γ_1 and γ_2 = unit weight of concrete and steel, respectively, kg/m^3,
V_1 = volume of concrete heated simultaneously in m^3,
δ_1 = thickness of the formwork,
F = area of the surface of incident infrared radiation in m^2,
τ_1 = time during which temperature rises from t_3 to t_1,
t_1 = temperature of isothermic concrete heating in °C,
t_3 = initial temperature of concrete before heating in °C,
t_4 = ambient air temperature in °C,
F_1 and F_2 = area of the exposed surface and of the surface covered with formwork in a structure in m^2,
k = coefficient of heat transfer from concrete through the formwork to the environment in $Kcal/m^2$ × hr × °C,
α = total coefficient of heat transfer by convection and radiation from exposed concrete surfaces of structures being heated in $Kcal/m^2$ × hr × °C
Ц = cement content in kg/m^3,
E_1 and E_2 = specific heat evolution of cement during the temperature rise and isothermic heating, respectively, Kcal/kg.

TABLE 6.3
Total coefficient of heat transfer by convection and radiation α

Value of k in	No wind	Moderate wind	Strong wind
	Wind velocity in m/sec		
$Kcal/m^2 \times hr \times °C$	up to 0.2	0.2 to 6	over 6
Up to 2.5	1	1.2	1.3
Over 2.5	1	1.3	1.4

The values of the heat transfer coefficients k are 4.6 $Kcal/m^2 \times hr \times °C$ for a wooden formwork 2.5 cm thick, 1.47 $Kcal/m^2 \times hr \times °C$ for concrete covered with a mineral wool mat five cm thick on a layer of roofing felt or roofing paper, and 2.1 $Kcal/m^2 \times hr \times °C$ for a wooden formwork 2.5 cm thick insulated with mineral wool five cm thick on a layer of roofing felt.

The value of α can be found in Table 6.3 according to k and wind velocity.

Infrared radiators combined with reflectors, heat insulators, and supporters are infrared units that have different designs according to their purpose. The following considerations govern the design of infrared units:

- A reflector is best be made of aluminum that has a high reflectivity. When aluminum is not available, it can be replaced with iron sheets whose reflecting surface is coated with heat-resistant aluminum paint.
- The supporters of infrared radiators should be made of light metals. It is particularly important when the radiators have to be moved manually. The weight of a unit should not exceed 20 kg.
- The absorbing surface of the formwork, if the heating is carried out through it, should be coated with black varnish whose emissive factor is 0.98.
- When vertical structures are to be heated, to ensure uniform heating and taking into account convection as well as infrared radiation, the electrical power should be distributed as follows: If we divide the structure to be heated conventionally into three equal parts, 50% of installed power should fall on the bottom third of the total height, 30% on the middle third, and 20% on the top third.
- To exclude the fringe effect, when heat loss of concrete is greater on the edges of a structure, the distribution of the power should also take it into account, 50% of installed power being supplied to the edge, then 30%, and then 20% in the middle.

Infrared heating should take into account the rate of temperature rise in the concrete on whose surface heat is to be supplied. It should never exceed 20°C per hour in all cases to ensure high quality of the concrete and the long life of the reinforced concrete structure being erected. When heating exposed structural members, the concrete surface must always be covered with a vapor insulation layer to avoid a large heat loss and overdrying of the concrete surface layer. The best cover is a thin metal sheet or foil coated with black varnish. The surface of concrete should never cool down sharply. The temperature must decrease as slowly as possible and not faster than 20°C per hour.

The infrared heaters are convenient for heating rooms for workers or enclosures when structures are cast in place inside. It is particularly characteristic of tall structures, such as reinforced concrete smoke stacks, towers, cores of high rises, etc.

The above described types of low-temperature and infrared heating of concrete are used to supply heat to outside surfaces of reinforced concrete structures. However, when heat is transferred this way, heat loss to the environment more or less always occurs.

6.3 HEATING FORMWORKS COVERED WITH A CONDUCTIVE MATERIAL

The design of heating forms with electrically conductive covers differ greatly from the heating formworks where heaters are installed on the back of the deck. The conductive layer is placed in them on the side of the deck that is in contact with the concrete and is well insulated from it. The deck may be made of metal or wood (water-resistant plywood). A layer of electrically conductive polypropylene three mm thick reinforced with polymer pile fabric and protected from the working side with a layer of inert polypropylene 0.5 to one mm thick is laid on an electric insulator. Power is fed to the conductive material by strip electrodes made of brass mesh with openings between one and 2.5 mm and the wire diameter of 0.35 to 0.5 mm, their specific resistance being 5 or 6×10^{-6} ohm.m. The insulator is polyurethane with a density of 40 to 45 kg/m3 (Figure 6.8). The heat insulation layer in this design acts as a layer that ties the heating cover with the structural base of the formwork panel. Electrode power leads protrude from the end of the insulation layer to be connected to a power source.

Practical applications of the heater rigidly attached to the structural base of the formwork revealed that it is quite hard to restore in case of any damage. Another disadvantage is that the formwork is used also in summer when the heaters are useless but still wear out.

So a better practice is to use the heating formwork with detachable heaters made of conductive polyurethane. The heater and the insulator are enclosed into a wooden frame, are fabricated separately from the formwork and are attached to it with bolts installed in the edges of the frame for winter concreting. When damaged, the heater can be easily removed, replaced with a new one, and then repaired. So the formwork panel is not put out of service while the heater is repaired and the construction schedule is not disrupted.

When a structure is heated, the nature of heat exchange of concrete with the environment is very important because it is a factor in heat loss and power consumption for heating. The key parameter in this case is the coefficient of heat transfer through the formwork.

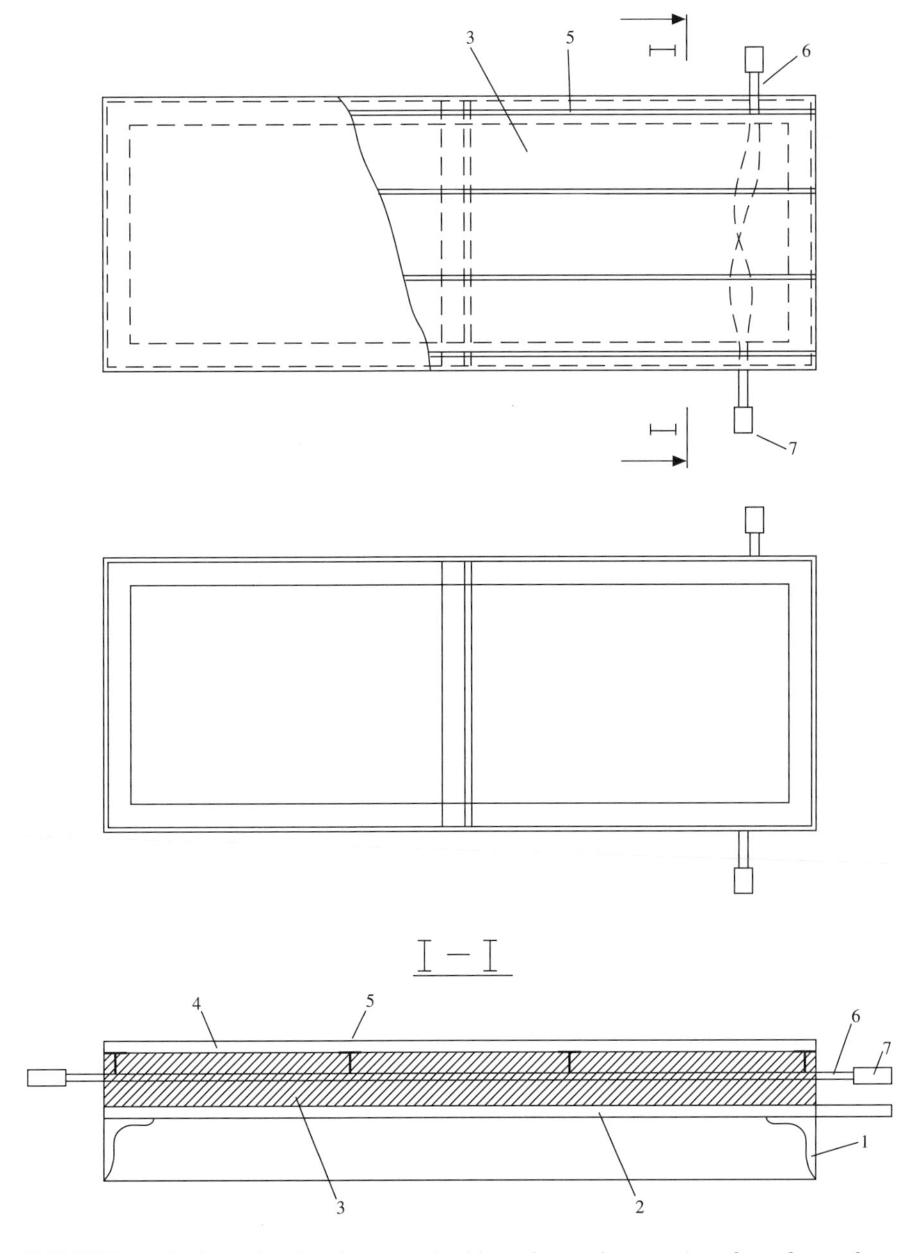

FIGURE 6.8 A thermal active form panel with stationary heaters. 1 — form frame, 2 — base of the form, 3 — foamed polyurethane insulation, 4 — heating cover, 5 — internal electric conductor, 6 — switching wires, 7 — couplings to switch panels.

Studies were conducted to determine it on heaters with foamed polyurethane heat insulators 20, 30, 40, and 50 mm thick (the coefficient of heat conductivity being 0.033 W/m^2 × °C) and a layer of polymer pile fabric placed on a formwork

TABLE 6.4
Values of the total coefficient of heat transfer (Ko) of a steel heating formwork with a foamed polyurethane insulator

	K_0 in W/m^2 × °C for insulator thickness in mm			
Wind velocity	20	30	40	50
0	1.121	0.827	0.655	0.542
1	1.130	0.832	0.658	0.544
3	1.441	0.989	0.753	0.607
5	1.504	1.019	0.770	0.618
10	1.522	1.026	0.774	0.621
15	1.538	1.034	0.779	0.624

made of metal sheets three mm thick. The steady-state heating conditions set after 12, 17, 15, and 37 hours. Averaged values of the coefficient of heat transfer at different wind velocities are given in Table 6.4.

Data produced by experiments do not differ practically from those found by calculation.

To minimize heat loss through the formwork so that the loss does not exceed 10%, the thickness of the insulation layer should be 21.9 mm. It is rather difficult to install a heat insulator of this thickness in practice and so it may be increased to 30 mm, which will only enhance its heat insulation properties.

The adhesion of the formwork to concrete becomes quite important when heaters made of conductive polypropylene are used. It affects not only the quality of the surface of a reinforced concrete structure being erected but also the life of the formwork, i.e., the number of its contacts with concrete.

The studies found that the surface of the conductive-polypropylene formwork with a protective cover has a much lower adhesion to concrete than that made of other materials. Thus, after first two turnovers of the formwork normal adhesion of the conductive polypropylene with a protective cover was 0.001 MPa while the adhesion of a steel deck was 0.65 MPa and of a plywood deck 0.081 MPa.

The adhesion to concrete increases as the number of turnovers of the formwork grows but differently for different materials. Of course, the preparation of the formwork to concreting is of great importance. After 20 turnovers the adhesion of the conductive polypropylene formwork rose to 0.037 MPa, of the steel formwork to 0.079 MPa, and of the water-resistant plywood formwork to 0.091 MPa. Although the adhesion of conductive polypropylene to concrete increased substantially as compared with its initial value and faster than that of the formworks made of steel or plywood, it still remained much lower than the latter.

The reliable heat insulation of foamed polyurethane in combination with conductive polypropylene makes concrete cool slowly even if the formwork is made of metal. Even at a temperature of –30°C it is possible in most cases to apply the least power consuming heating procedure called "electric thermos" when placed concrete

TABLE 6.5
Recommended temperature rise rates on the heated surface

Thickness of structure in m	0.05	0.10	0.20	0.30	0.40	0.50
Heating rate in °C/hr	15	13	6	4	2	1.5

TABLE 6.6
Recommended maximum temperatures of heating the surfaces of structures

Thickness of structure in m	Ambient temperature in °C	Maximum temperature of heating surface of structure in °C when concrete attains required strength by end of cooling in % R_{28}				
		30	40	50	70	80
0.1	–10	20	20	30	50	70
	–20	20	30	40	55	70
	–30	30	40	45	60	75
0.2	–10	20	20	30	40	50
	–20	20	30	30	40	55
	–30	30	30	35	40	60
0.3	–10	20	20	25	35	45
	–20	20	25	30	40	50
	–30	25	30	35	40	50
0.4	–10	15	15	20	30	40
	–20	20	20	25	35	40
	–30	20	25	30	40	40
0.5	–10	15	15	20	20	35
	–20	15	15	20	25	40
	–30	15	15	20	30	40

is heated to the required temperature whereupon the current supply to the electrodes is cut off. Due to small heat loss to the environment and hydration heat evolution as the concrete hardens the latter cools very slowly and thus can harden quite rapidly. The rate of temperature rise should be different according to the thickness of the structure being heated and the heating rate is lower for more massive structures (Table 6.5).

The application of the electric thermos procedure depends on the level of maximum temperatures on the heating surface of the formwork according to the ambient temperature and its massiveness (Table 6.6).

For structures thicker than 0.5 m the temperature on the concrete surface should rise slowly and not exceed the temperature in the core. It must be maintained in the surface layers at a lower level as the core is heated by cement heat evolution to avoid high internal stresses in the structure and cracking at the surface. In other words, a favorable thermal stress state typical of the thermos curing will be created in the structure.

Uniformity of heating. Heating of a cast-in-place structure in a heating formwork with any type of heater always requires that a uniform temperature field be provided over the whole cross section of the concrete. The situation becomes more difficult for structures cast of lightweight concrete that has a lower density than normal weight concrete and thus lower heat conductivity. It can be seen very clearly in heating cast-in-place structures of different thicknesses made of agglоporous concrete with a unit weight of 1560 kg/m^3.

Fragments of cast-in-place structures 150, 200, 300, 400, and 500 mm thick were heated at an above-zero ambient air temperature in the range of 9 to 14°C and at a subzero temperature that varied from –12 to –16°C. The heating formwork was a metal deck four mm thick provided on the back with strip heaters made of a conductive fabric enclosed in an electric insulator made of several layers of glass-fiber cloth. The formwork was insulated outside with a 50 mm layer of mineral wool. An analysis of the studies proved that the temperature field in the concrete was clearly nonuniform even at an above-zero ambient temperature. Thus, as fragments 150, 200, and 300 mm thick were heated at a rate of 20°C/hr, the temperature difference between the layer of the aggloporous concrete in contact with the thermally active panel and the core of the fragment increased with the level of isothermal curing that rose from 40 to 80°C. After the temperature of isothermal curing was reached at the interface of the concrete with the thermally active panel the heat flux became more uniform. From that moment on the difference of temperatures between peripheral layers of the concrete and its core decreased as the curing continued and gradually leveled off.

An even more striking picture was observed in electric heating of fragments 400 and 500 mm thick. When temperature grew at a rate of 15°C per hour, its maximum difference between the layer of the aggloporous concrete in contact with the thermally active panel and the core increased with the temperature to between 25 and 36°C when the thickness was 400 mm and 38 to 46°C for a thickness of 500 mm by the start of isothermal curing. The temperatures practically leveled off by the end of the curing.

Should the temperature of isothermal curing drop below 70°C, the leveling off of the temperature over the cross section becomes practically impossible. It should be emphasized, however, that in thick structures the temperature in the core may non only become equal with the surface layers but even exceed it due to added heat of hydration.

The experimental studies demonstrated that due to the thermal treatment, when the heating rate exceeded 20°C/hr, the structure of the aggloporous concrete was dense, uniform, and close to that of the concrete that hardened under normal conditions of humidity and temperature.

The formation of temperature fields in similar fragments at a subzero ambient temperature was different only at an early stage, i.e., immediately after the concrete was placed. In this case the temperature of the concrete placed in a cold form dropped in the contact layer below the temperature in the core because of the heat transfer to the cold formwork. However, only one hour after the heating started the temperature leveled off over the cross section, then rose faster than that in the core and leveled off again at the end of the isothermal curing.

An analysis of these studies of temperature fields formed in heating aggloporous concrete disclosed that their formation is roughly identical at both above-zero and subzero ambient temperatures. The heating time increased somewhat at a subzero ambient temperature because of a low above-zero temperature of the concrete at the interface with the formwork, which increased the time of the thermal treatment. This increased electric power consumption by 24 to 27% as compared with the thermal treatment when the ambient temperature was above zero.

6.4 APPLICATION OF ELECTRIC CONTACT HEATING OF CONCRETE

Electric contact heating of concrete is used in many countries for construction in cold weather. In Finland, walls of buildings are cast in place in heating formworks made of water-resistant plywood with heating wires. Heating formworks with mesh heaters were used in France in erecting residential buildings of cast-in-place reinforced concrete (internal walls and floor slabs). Heating wires have been also used to heat concrete in structures with large exposed surfaces. String heaters are used in Russia to heat concrete in joints between precast reinforced concrete columns. Heating wires were employed in Moscow in building a trade center on Manezh Square to heat concrete in cast-in-place columns and floors (Figure 6.9). The heating formwork with mesh heaters was used in construction of residential buildings. Fabric heaters have been used in Belorussia.

The multitude of types of electric heating of concrete is an evidence of the great potential of the method. We can always choose the most convenient procedure of electric heating with low-temperature or high-temperature heaters in erecting any cast-in-place structure. It allows to carry out cast-in-place reinforced concrete construction in cold weather faster than in summer when concrete is cured in natural conditions until it attains the required strength.

So it is no accident that concrete may be subjected to electric thermal treatment also in the summertime to accelerate hardening and the construction in general. The electric contact heating is the most common method in those cases.

Since we deal there with electric current, safety precautions must be always taken when concrete is heated electrically. Care must be taken to secure good contacts between heating elements and between heaters and power leads. Electric insulation of the heating elements from the formwork should be closely watched. As a heating formwork is installed, the safety of the electric insulation of heaters and of all connections should be checked before the heat insulator and the protective cover are placed whereupon the formwork can be completed and put into service.

The condition of the electric insulation of power leads should be closely monitored at all times during operation. These should be installed so as to avoid any damage to them. When formwork panels are deenergized and moved to another place, their contacts should be protected against damage and contamination. All operations of installation and dismantling of formwork and repairs of heating elements must be always performed with power cut off.

FIGURE 6.9 The reinforcing cage of a column with a heating wire in it (white color) prepared for installation of the formwork and concreting at a construction site in Manezh Square in Moscow (December 1995).

When work is to be done at night, the electric heating sites must be well illuminated and red lamps must be installed at the boundaries of the heating area to warn that electric thermal treatment is in progress.

7 Electrode Heating of Concrete

Electrode heating is a most effective method used to accelerate the hardening of concrete. With this method, the concrete is brought into electric circuit after being placed and heats up just like any resistor as electric current is passed through it. Its temperature rises accelerating the reaction of cement with water and thus the hardening of the concrete. Electric power is used more efficiently since it is converted to heat in the concrete itself.

Early studies of the application of electric current to concrete heating by making the material part of a circuit were started in 1924 by Jones, Fleming, and Tagge at the Montreal laboratory of the Canadian Cement Company. After experiments with concretes containing admixtures of calcium chloride and aluminum chloride the researchers came to the conclusion that electric current increased the compressive strength of concrete. Curiously, they did not attach any importance to the factor of temperature, but tried to explain the strength gain by the effect of electric current that "helped to disperse cement particles more evenly and perfectly over the mass of concrete." Although concretes and mortars were never heated during the experiments higher than 33 to 45°C, they were undoubtedly affected mostly by the temperature factor and it was a mistake to ascribe the role of a stimulator of uniform distribution of cement grains in heated concrete to electric current.

The erroneous theoretical concept of electric heating as applied to concrete did not allow the researchers to see practical benefits of the method and delayed publication of their experimental results until 1936, i.e., four years after Swedes had received a U.S. patent.

An outstanding part in the actual discovery of the electrode heating of concrete was played by the Swedish engineers Albert Brund and Helge Bolin. Having started their investigations in 1927 in the Hernisand Gymnasium of Technology, they not only put forward the correct scientific concept of the method but also appreciated its practical potential by demonstrating the commercial application of electrode heating to casting reinforced concrete structures. They held that when AC current passed through fresh cement mortar or concrete mixture, the material released heat which accelerated reactions between water and cement in the mixture. In 1932 the inventors received a U.S. patent, published their data and made a report at the Swedish Engineering Academy. That was the beginning of the electrode heating of concrete that started to be used at construction sites in winter, in the Soviet Union in the first place. As early as February and March 1933, under the guidance of A. K. Reti, a Soviet engineer, the method was used in the construction of an industrial building near Moscow to accelerate the hardening of concrete in columns, wall beams, and crane beams. And that was not just an isolated experiment but a procedure on a commercial scale that produced 10 to 12 cu.m of reinforced concrete per day with a power consumption of 80 kW/m^3.

Since that time the Soviet Union became the center of research and wide application of electrode heating of concrete in construction bearing the palm to this day. The transfer of the center of research into electrode heating as well as into other methods of electric thermal treatment of concrete to the Soviet Union was not accidental. Construction projects were underway all over the country on a gigantic scale and it was impossible to stop them in winter. Cold weather lasts in this country for four to eight months and it would be disastrous economically to stop the construction for so long. Buildings have been erected and are erected there now the whole year round nearly at the same pace.

Electrode heating has been developed and still is being developed now in Russia. A. K. Reti, R. V. Vegener, S. A. Mironov, A. S. Arbeniev, V. Ya. Gendin, and many other Soviet researchers made great contributions to the theory and practice of this method.

Today several millions of cu.m of concrete are placed at winter construction sites in Russia every year maintaining a high pace of erecting buildings and engineering structures at any ambient temperature.

7.1 ELECTRIC CONDUCTIVITY OF CONCRETE

Electric conductivity or its reciprocal — resistance — is a basic characteristic of any material that is brought into circuit as one of its links. Concrete plays the part of this link in electrode heating. As current passes through it, the electric power is transformed into heat and the conductor heats up. Having ion type of conductivity, concrete is characterized by resistance that is unstable with time due to physical and chemical processes in it.

Many factors, including the chemical and mineralogical composition of cement, the presence of admixtures, the heating temperature, and water–cement ratio, affect the multicomponent system such as concrete. These factors put together may so much affect conductivity that concrete may either not heat under effects of designed electric characteristics or, on the contrary, heat up so fast that the heating conditions are upset and the electric equipment is overloaded. This is the reason why so much emphasis was laid in research on studying the electric conductivity of concrete.

Studies disclosed that concrete had three main components on which electric conductivity depended. These are cement, water, and admixtures.

Dense aggregates whose solubility in water is negligible have a very high electric resistance and are of no account. Dry porous aggregates are dielectrics but their specific resistance drops dramatically when they are saturated with water. Still it is measured in many thousands of ohm.cm. Since it is many times higher than the specific resistance of cement hardened paste and of the liquid phase, we can also neglect this factor.

Due to its high water-retaining capacity and solubility of clinker minerals, cement has rather high conductivity when mixed with water and can be called an electron conductor. Dry cement is a dielectric.

Water is the main component on the amount and chemical composition of which the electric conductivity of concrete depends at various hardening stages. An absolutely pure water has an enormous specific resistance (several millions of ohm.cm) and is an excellent dielectric. However, even a negligible content of impurities changes the picture sharply. Clean, good-quality tap water, for example, has a specific resistance already measured in eight to 12 thousand ohm.cm.

After concrete is mixed with water and agitated the chemical composition of the latter changes drastically due to dissolution of cement components and the water becomes a good electric conductor. Chemical admixtures added to concrete for different purposes also help greatly. In view of high conductive properties of water, it is considered to be the basic component in a concrete mixture, which makes concrete electrically conductive. Henceforth we call water with dissolved cement clinker minerals the liquid phase and use largely this term.

Investigations of the electric conductivity of concrete in hardening started early in the 20th century. H. Gessner and I. Shimitsu, N. A. Moshchansky and Calley, C. Dorsh, A. Saul, Yu. S. Malinin, V. E. Leirikh, V. Ya. Gendin, G. Kalouzek, R. V. Vegener, P. Huller, Ya. Ichiki, and many other researchers studied various factors affecting the electric conductivity of concrete. Their views often differ as far as the importance of a particular factor is concerned but the experience accumulated with time helped to make the state-of-the-art interpretation of this phenomenon on the basis of extensive studies conducted at the NIIZhB. This paper intended for practical use by engineers and technicians of design offices and construction contractors cannot analyze all theoretical aspects of the electric conductivity of concrete. So we cover this problem only briefly and in an easily understood form without analyzing views of various researchers.

The electric conductivity of concrete is a function of variation in the qualitative and quantitative characteristics of the current-conducting component, which is the liquid phase. The variation in the specific resistance of concrete with time can be divided into three distinct stages, including a drop of ρ to an extreme value, stabilization at values close to the extremum, and an increase as the concrete hardens. Each stage reflects a complex set of physico-chemical processes that go on in concrete as it hardens under different temperature conditions. Consider the relationship between the variation in the electric conductivity of concrete and the processes in it at different stages.

Mixing of concrete with water triggers a rapid dissolution of caustic alkalis and lime and saturation of the liquid phase. The alkaline components contained in most cements are dissolved quickly after the concrete is mixed with water. We may safely say that most of them become part of the liquid phase within first 15 to 60 minutes.

As temperature rises, the process of dissolution accelerates due to the increased reactivity of water and the saturation of the liquid phase by the alkalis takes less time. The dissolution of CaO and the formation of $Ca(OH)_2$ in the liquid phase lasts longer and depends on the composition of cement. We can say that the concentration of $Ca(OH)_2$ in the liquid phase reaches its maximum within three to six hours after mixing. The solubility of lime in the presence of potassium and sodium alkalis drops because of the presence of the same hydroxyl ions OH^-.

The sharp drop of ρ after mixing is caused by fast dissolution of KOH and NaOH. As less and less alkalis of potassium and sodium get into the solution, the curve smoothes out and gradually reaches its extremum value. Thus, in experiments with a suspension, ρ changed by 26 to 35% in the first three hours as compared with its initial value and by only 7 to 10% in the second three hours. This time and variation ranges are somewhat higher in concrete.

At this stage the role of calcium hydroxide starts to be felt to a greater extent, its concentration increasing in the liquid phase to oversaturation. The electric conductivity of concrete reaches its maximum in three or four hours after mixing.

Along with the qualitative variation in the liquid phase, its quantity also changes because the water is bound chemically in new formations, adsorbed by the surfaces of microcrystals of the new formations, and evaporates to the environment. The change in the quantity of moisture is not great in the first two hours and does not exceed 1 or 2%. Further on the process goes faster. Under heating the moisture loss by evaporation may prevail. As the electric conductivity of concrete approaches its maximum value, the quantitative aspect of the liquid phase becomes increasingly important while the qualitative aspect means less and less.

At the extreme point, which may be quite distinct in heating or become an area of rather long stabilization at low above-zero temperatures, there is a balance between two processes that are opposite in their effects on ρ.

Having reached an extremum, the curve of ρ starts to rise slowly at first and then much faster. The main reason for the change in the specific resistance of concrete at this stage is the reduction of the amount of mechanically bound water which is the main conductor of electric current in concrete. The key role of mechanically bound water in the characteristic of the electric conductivity of concrete is proved by the variation in ρ in freezing. In this case concrete becomes a dielectric rather than a conductor because its specific resistance may be as high as several tens of thousands of ohm.cm at a low temperature. It is impossible to heat frozen concrete using voltages that are commonly applied for this purpose.

The effects of cement content on the specific resistance of concrete. The content of cement affects primarily the qualitative composition of the liquid phase and its effects could be seen quite well in a study of cement-water suspensions containing the same amount of water but a variable amount of cement (from 20 to 500 kg when scaled to one cu.m of concrete).

The experiments found that even a small amount of cement in water reduced its specific resistance dramatically. The latter continued to drop as the cement content increased to 120 kg/m^3 and almost did not change when it was increased further. A similar situation was found in suspensions and cement paste using different types of cement.

As can be seen from the above data, the cement content between 100 and 450 kg/m^3 changes specific resistance by no more than 10%. It gives us ground to believe that where the water content is constant (200 l per 1m^3 of concrete), the specific resistance of cement paste or suspension may be assumed to be practically constant.

This phenomenon has a good explanation in terms of physico-chemical processes that take place when cement is mixed with water. Addition of even a small amount of cement to water increases its electric conductivity many times for a very short period counted in minutes. Caustic alkalis play a leading part there, the concentration of ions K^+, Na^+ and of hydroxyl OH^- increasing quickly with dissolution.

The physico-mechanical factor becomes more important when the cement content increases. Namely, the higher cement content in the solution increases the amount of electrolyte and at the same time the increase in the dispersed fraction reduces the amount of mechanically bound water since it is adsorbed by cement grains and thus the cross section of the current conducting component.

The effects of the physico-mechanical factor and of the physico-chemical factor are balanced when the cement content is about 200 kg/m^3. With higher cement contents the former tends to prevail. It can be confirmed by the nature of change in the electric conductivity of the filtered liquid phase that continues to increase with the cement content. It should be noted, however, that the change in electric conductivity is not proportional to the cement content. Thus, when the latter increases by a factor of six (from 50 to 300 grams per 500 cm^3) the specific resistance of the liquid phase drops 2.0 or 2.5 times and with an increase in the cement content from 300 to 1000 grams, i.e., by a factor of 3.3, for the same amount of water, specific resistance decreases only by 30 to 35%.

The disproportionality in the variation in the specific resistance of the liquid phase as the cement content increased can be explained in terms of electrochemsitry. The electric conductivity of electrolyte is a derivative of the concentration of ions and of their mobility. Higher concentration increases the electric conductivity of electrolyte to a certain limit. Since strong electrolytes, which include KOH, NaOH and $Ca(OH)_2$, practically disintegrate into ions in a solution,* the slower increase in the electric conductivity can be caused only by a drop in mobility of ions in the solution as its concentration rises. It is quite in order because each ion gets surrounded after disassociation by opposite ions that form a so-called "ion cloud." Water molecules in the space between the ions form, in addition, solvation shells around them, which prevent the ions from joining back into molecules.

The imposition of an external electric field stimulates the movement of ions to opposite electrons but the solvation shells and the ion cloud inhibit it considerably. Collisions of ions, whose number increases with concentration, retard their movement even further.

All these phenomena reduce electric conductivity. So an increase in the cement content increases the number of ions but decreases their mobility and thus opposite effects on ρ compensate each other. This is the reason why the cement content in the above range affects the variation in ρ very little.

The effects of the cement content on (in concrete are somewhat different. A comparison of conductive properties of concretes containing the same amount of water but different amounts of cement (and thus different quantities of aggregates) revealed a certain relationship between variation in ρ and the amount of the binder. As the cement content increases, the initial and minimum specific resistance of concrete decreases (Figure 7.1). Thus, with the cement content rising from 300 to 500 kg/m^3, the specific resistance of the concrete dropped by 14 to 17%. The reason was an increase in the quantity of the current conducting component — cement paste — in the concrete and improvement, though small, in the latter's electric conductivity. This conclusion was confirmed by the time taken by the concrete to reach minimum specific resistance (rmin). It was about 1.3 times shorter for concrete containing 500 kg/m^3 than for concrete with a cement content of 250 kg/m^3. An

* This assertion does not refer to highly concentrated electrolyte solutions

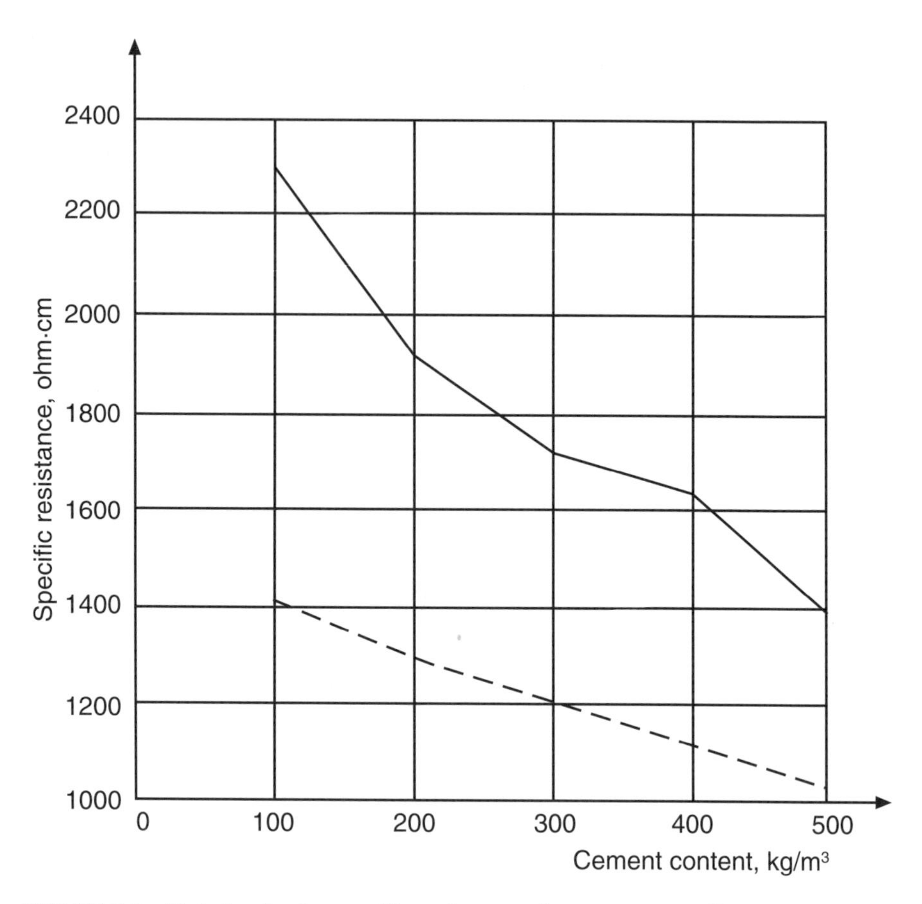

FIGURE 7.1 Variation in the specific resistance of concrete according to its cement content. ________ initial specific resistance, ---- minimum specific resistance

important factor in this case appears to be the rate at which the liquid phase was saturated with calcium hydroxide and free alkalis, whose content increased with the cement content while the water content remained constant.

The effects of water content on the specific resistance of concrete. As was mentioned above, water affects the electric conductivity of concrete most of all. Thus, when the water content is brought up to 325 from 135 l per $1m^3$ of concrete, the initial and minimum specific resistance of the concrete decreases (Figure 7.2). When concrete hardens under normal conditions, the period of quazistabilization of the specific resistance value close to minimum lengthens considerably with an increase in the water content. Growth of the water content slows down the increase in ρ after the minimum was reached. It is quite natural because chemical binding of water in new formations and loss of water due to evaporation and adsorption by crystals of the new formations affect the total balance of water, particularly mechanically bound one, when the amount of mixing water was small at the start.

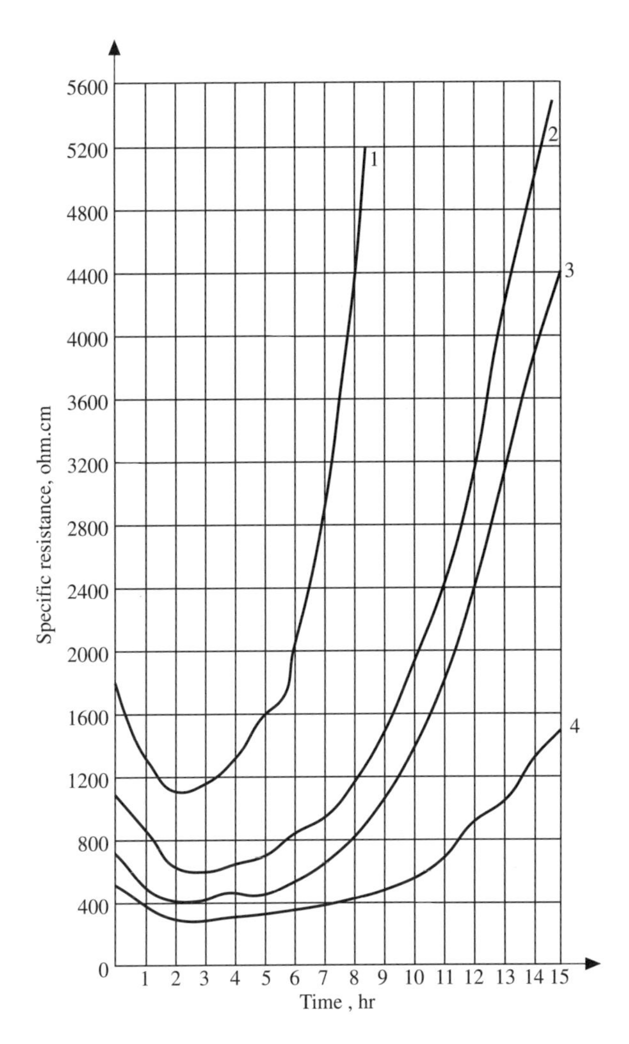

FIGURE 7.2 Variation in the specific resistance of concretes of different compositions heated at 80°C using the mode of 5 + 8 + 4 hours. 1 — cement: 300 kg/m^3, water: 135 l/m^3, 2 — cement: 300 kg/m^3, water: 195 l/m^3, 3 — cement: 500 kg/m^3, water: 225 l/m^3, 4 — cement: 500 kg/m^3, water: 325 l/m^3.

The initial and minimum values of specific resistance are of paramount importance for the nature of its variation in concrete. Experiments found that with any water content up to 200 or 220 l/m^3 specific resistance decreased at about the same rate of 15 ohm.cm/l. When the water content was higher, the rate may drop as low as two ohm.cm/l. The reason should be sought in the structural changes of the material in hardening.

When the content of water is low, it is bound almost entirely and with a different strength physically. So there is little mechanically bound water left. This disrupts the continuity of the current conducting channels and increases ρ. As the water content increases, the amount of mechanically bound water grows and thus the continuity of the capillaries filled with the conductive component improves. The higher amount of water in concrete decreases the difference in the initial and minimum values of ρ, which should be taken into account at the site in electric heating of concrete. Taking the ratio of minimum ρ to its initial value, we can say that it is about 0.6 to 0.74 in concretes containing 135 l of water per 1 cu.m and 0.64 to 0.83 in concretes containing 325 l/m^3.

In approximate calculations to determine electric power required to heat concrete, the mean ratio of the minimum specific resistance to its initial value can be assumed to be 0.7.

The effects of temperature on the variation in the specific resistance of concrete. The kinetics of the variation in the electric conductivity of concrete in the process of heating and in freezing changes considerably as compared with normal temperature conditions.

Studies have shown that ρ changes in the same way decreasing first to an extremum and then starting to rise. The difference lies in that these stages become shorter since the physico-chemical processes go faster.

Comparing the variation in ρ in concretes heated to 60 and 80°C (Figure 7.3), we can see that the time required to reach minimum ρ is almost no different and the stabilization period differ but little from normal conditions. However, the rate of change in ρ along the branch of the curve rising after the extremum increases with temperature. The difference in rates is 13% after 10 hours of heating, 63% after 12 hours, and 95% after 15 hours of heating, or in absolute terms the difference within the same period is 1.8, 2.3, and 3.1 times, respectively.

The variation of ρ with temperature is affected to a greater extent by the concentration of electrolytes in the liquid medium and by the mobility of ions.

The solubility of calcium hydroxide is known to drop with rise in temperature. So the saturation concentration of 1000 g of water is 1.3 g of CaO at 0°C, 1.18 g at 20°C, 1.00 at 40°C, 0.82 at 60°C, 0.658 at 80°C, and 0.523 g at 99°C. Thus, the solubility of lime does not increase electric conductivity with temperature. Equivalent electric conductivity drops 1.8 times at 80°C as compared with 20°C due to a lower concentration of $Ca(OH)_2$ in the solution. The situation is quite different with the mobility of $Ca(OH)_2$ ions — it rises fast with temperature. So Ca^{++} ions are four times and OH^- — 1.6 times as mobile at 80°C as at 20°C. The equivalent electric conductivity increases 2.1 times. So the mobility of ions is a key factor in increasing electric conductivity with temperature.

The reason for a little lower minimum value of ρ in the liquid phase with some cement must be a higher content of caustic alkalis in clinker.

In addition to the heating temperature, the time needed to reach the minimum specific resistance is also affected by the rate at which the temperature rose, which may vary over a wide range.

Studies found that with a relatively slow heating rate the minimum ρ in normal weight concretes can be reached a little earlier than the temperature of isothermal curing. Lightweight concrete with expanded clay heated at the same rate reaches ρ by the time the maximum heating temperature is reached. It happens in the former because of the predominant influence of the chemical and physical binding of water

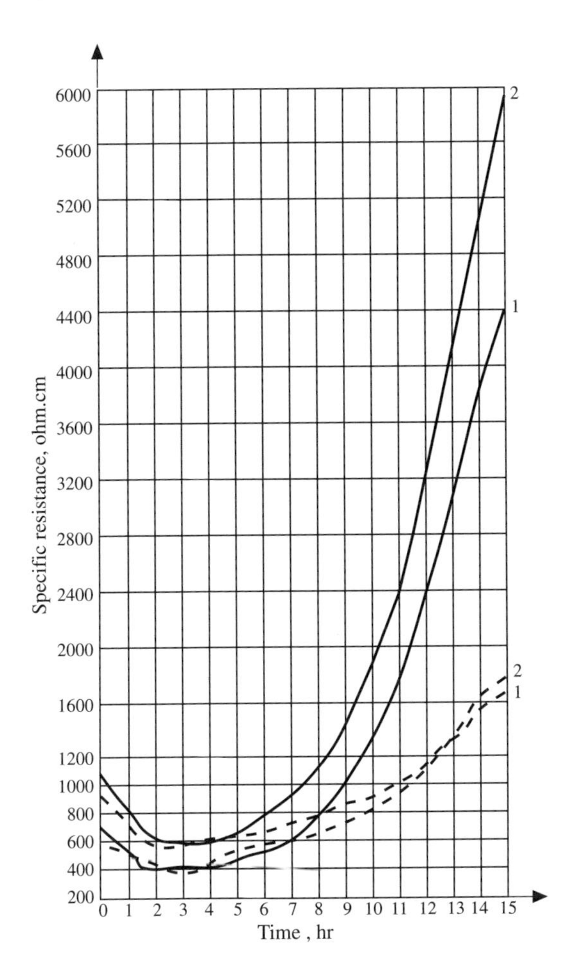

FIGURE 7.3 Variation in the specific resistance of electrically heated concretes of different compositions. 1 — concrete containing 500 kg/m^3 of cement and 225 l/m^3 of water, 2 — concrete containing 300 kg/m^3 of cement and 195 l/m^3 of water, ________ heating at 80°C, ---- heating at 60°C.

in new formations. In the latter it happens due to adsorption of moisture in the concrete mixture by porous aggregate. As the temperature rises, the air in the aggregate pores expands and forces water out into the space between particles. This increases the amount of the liquid phase although the new formations appear and water is bound chemically in lightweight concretes at the same rate as in normal weight concretes. When the maximum temperature is reached and stabilizes during isothermal curing, the expansion of the air stops, moisture no longer comes out of the aggregate and electric conductivity starts to increase quickly because the water continues to be bound in new formations and evaporates.

The minimum ρ may be reached by the end of heating in normal weight concrete too as the temperature rises at a rate of 60°C per hour or faster when it depends mostly on the saturation of the liquid phase and an increase in the mobility of ions (Figure 7.4).

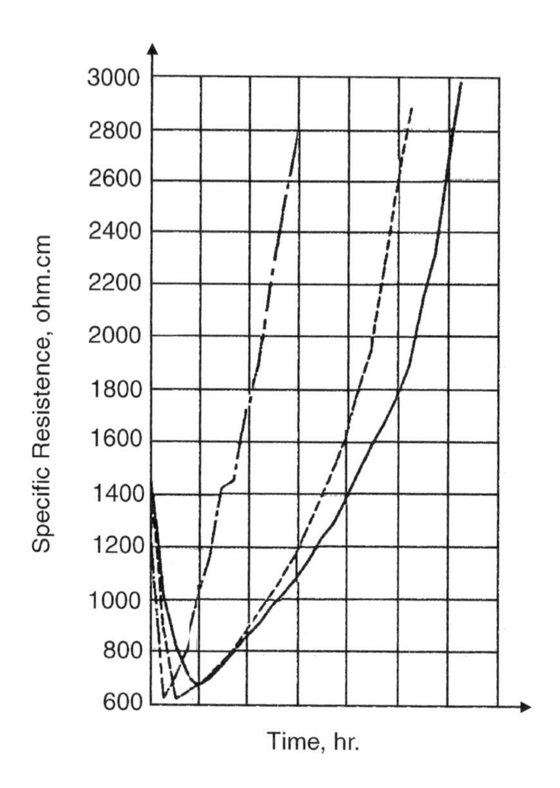

FIGURE 7.4 Variation in the specific resistance of concrete heated to 95°C at different rates. _________ heating at the rate of 80°C/hr, ---- heating at the rate of 160°C/hr, __.__.__._ heating at the rate of 320°C/hr

The minimum specific resistance of concretes heated at different rates within 80 to 320°C does not change much and the fluctuations do not exceed 7%. But in concretes heated at a rate of 500°C the difference in minimum values of ρ may be as high as 16%.

Variation of specific resistance at 0°C or lower has a great practical importance for electric heating of concrete in winter.

Studies have shown that specific resistance increases with a drop of temperature from 20 to 0°C (Figure 7.5). The increase in ρ accelerates when concrete starts to freeze and the singular point can be clearly seen on the curve. Then ρ increases evenly and very fast as the concrete continues to freeze.

The reason for lower electric conductivity when concrete is cooled to 0°C is a lower mobility of ions when concrete is cooled to 0°C and phase changes of water when it freezes since the specific resistance of ice is very high being 166×10^5 ohm.cm.

The rate of change in ρ at low above-zero temperatures depends on the amount of water contained in concrete — it drops when the water content increases (Figure 7.6). At subzero temperatures ρ remained practically constant in all fresh concretes that were studied. At a temperature of –1 to –3°C the specific resistance of

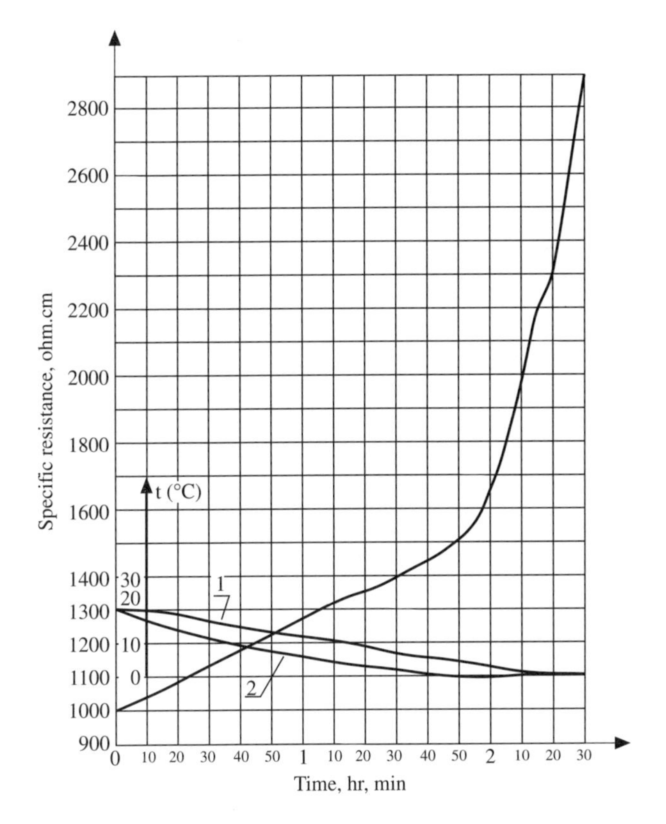

FIGURE 7.5 Variation in the specific resistance of concrete in the process of cooling and freezing. 1 — temperature of concrete in the core of a specimen, 2 — temperature of concrete on the surface of a specimen.

concrete ranges from 15 to 30 thousand ohm.cm (according to water content) and so the concrete cannot be warmed up using voltages generally used in electric heating.

So concrete must never freeze where it contacts an electrode before electric heating starts.

To sum up we can say that the electric conductivity of concrete is above all a function of the amount of mechanically bound water it contains and of the way the latter is saturated with electrolytes. These main factors are the reason for the variation of electric conductivity with time due to relaxation of the material in hardening. Temperature may accelerate or retard the kinetics of the process while the general nature of variation in ρ remains constant.

7.2 PROCESSES IN CONCRETE CAUSED BY ELECTRIC CURRENT PASSING THROUGH IT

Concrete is generally heated with AC current of industrial frequency. Regrettably, direct current is sometimes promoted as a source of heat in concrete. Attempts are

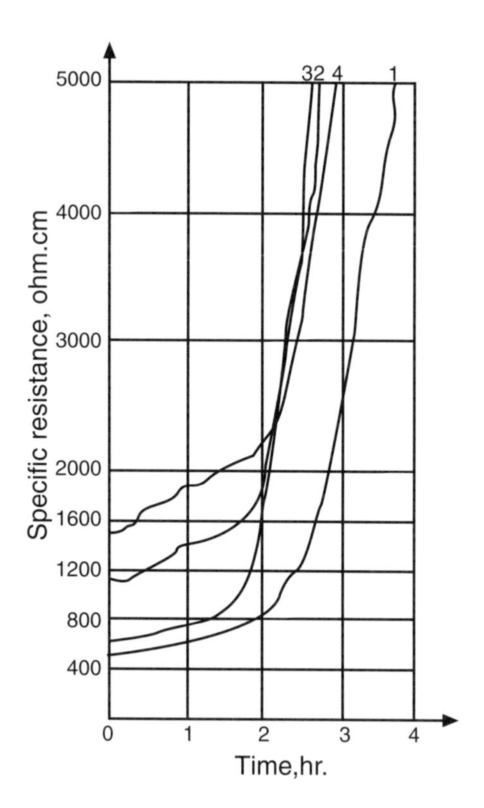

FIGURE 7.6 Variation of the specific resistance of concretes of different compositions in the process of freezing. 1 — concrete containing 500 kg/m^3 of cement and 325 l/m^3 of water, 2 — concrete containing 500 kg/m^3 of cement and 225 l/m^3 of water, 3 — concrete containing 300 kg/m^3 of cement and 195 l/m^3 of water, 4 — concrete containing 300 kg/m^3 of cement and 135 l/m^3 of water.

made to justify this idea theoretically. Consider in brief the phenomena that occur in concrete when electric current is passed through it.

The inclusion of an aqueous solution of electrolyte (the liquid phase of concrete) into the DC circuit gives a direction to the chaotic motion of ions at once. The higher is the gradient of the external electric field the better the directivity of the motion. The rate of the processes depends on the speed of ions migrating towards electrodes. For cations, it is 0.2 cm/min for H^+ and 0.027 cm/min for Na^+, and for anions, 0.11 cm/min for OH^-, 0.0415 cm/min for Cl^-, and 0.039 cm/min for O_3^- at 18°C. Notice that the speed of hydrogen and hydroxyl ions is higher than that of all other ions by an order of magnitude, the reason being continuous reactions of hydrogen ion exchange between hydroxonium ions, H_3O^-, and water molecules and between water molecules and hydroxyl ions. So it is those ions that are the most active and capable of carrying electric charges faster than others (i.e., give away or catch electrons on electrodes).

When a DC field is applied to concrete, three phenomena largely occur there. They are electrolysis, electrophoresis, and electroosmosis. There are also some processes on electrodes, the basic ones being covered below as applied to concrete. The assertions that DC current may have a favorable effect on the structure and properties of concrete were not confirmed.

The electrolysis of water is the most distinct and noticeable phenomenon observed during DC heating of concrete. It manifests itself as gaseous hydrogen released on the cathode and oxygen on the anode. The former is produced when hydrogen ions are discharged and join into molecules. The latter is the product of the discharge of hydroxyl ions followed by formation of electrically neutral OH groups that cannot exist by themselves and change into molecules of water and oxygen. The gases are released on the electrodes quite rapidly and can be seen quite well visually.

Studies found that in electric heating of concrete the electrolysis has adverse effects producing mainly two phenomena — polarization and oxidation of electrodes.

Polarization increases contact resistance between an electrode and concrete due to the release of gases during the decomposition of water and consequently makes the electric heating process more difficult. Oxidized electrodes have a shorter life, involve additional costs because they need cleaning, leave spots of rust on the concrete surface, and certainly increase contact resistance since parts of the electrodes become shielded by corrosion products.

Reversing the polarity every two to five minutes, as some researchers recommended, makes things only worse, not better, because both the cathode and anode are subject to corrosion in this case.

Electrophoresis (catophoresis) is a migration of dispersed solid particles, gas bubbles, and colloidal particles caused by applied electric field. Our repeated attempts to determine by experiments the role of this factor in the process of electric heating of a cement suspension unfortunately failed. There was no noticeable migration of dispersed particles to electrodes under voltages from 20 to 220 V with the electrode spacing of 20 mm. Strength characteristics of concretes heated with DC and AC current differ but little in favor of the latter. So electrophoresis does not affect the structure and strength of concrete to any noticeable extent.

Electroosmosis, which is migration of the liquid phase from the anode to the cathode, does occur. But its role in electric heating of concrete is not great although it may be used to dehydrate concrete.

Thus, the accelerating effect of DC current on the hardening of concrete due to electrophoresis and electroosmosis phenomena that allegedly help cement grains to hydrate deeper because moving particles remove gel films from their surfaces is negligible and may not be taken into account.

Studies conducted by the NIIZhB on the effects of DC and AC current in electric heating of concrete proved that the former had no advantages over the latter but its adverse effects were obvious. It was confirmed many times in practice at the construction site.

The foregoing gives us reasons to believe that it makes no sense to use DC current in electric heating of concrete and that AC current should be applied instead. But then the question arises: does the extremal electric field applied to concrete affect its structuring processes or hardening directly or the concrete hardens faster only due to higher temperature?

Some researchers, including A. K. Reti, A. Fleming, O. P. Mchedlov-Petrosyan, and Yu. N. Vershinin, said that the electric current affected the structuring of concrete improving its strength. But other researchers, e.g., S. A. Mironov, P. I. Bazhenov, V. P. Ganin, and G. A. Polkovnikova did not find any noticeable influence of electric current on the structure of concrete in their experiments. This problem is covered in the literature insufficiently and erratically and the data presented in papers were mainly obtained phenomenologically rather than based on direct proof of the impact of electricity of the hardening process. Let us analyze, if only in brief, the hardening process in concrete under an applied electric field.

The cement-water system has a vast number of microcrytalline formations of colloidal or larger sizes. When a low-frequency electric field is applied, the microparticles may become oriented along force lines. Yu. N. Vershinin suggested for this reason that the electric thermal treatment of concrete be carried out in two stages with the electric field to be applied at the early stage immediately after placing of concrete for a short time and the second stage of thermal treatment to be started after an interval during which the structure should strengthen a little. Unfortunately, that proposal was never implemented in practice and there is no proof of its effectiveness.

We investigated effects of electric current on concrete hardening at the NIIZhB, including the method suggested by Yu. N. Vershinin. No appreciable influence on the strength of concrete was observed. There is good reason for it because the strength of concrete depends, in addition to hardened cement paste (whose content in concrete is lower than 30%), also on the strength of aggregates, the nature and magnitude of internal stresses, the reliability of the contact between the aggregate and the binder, etc. It seems plausible that the role of the electric field, even if it appreciably affected cement paste, was reduced to zero in concrete by actions of other, more tangible factors.

Nevertheless, it would be wrong to reject the idea of an influence of the electric field on structuring because the system is predisposed to such action. A certain effect may be produced by electrophoresis due to polarization of microparticles, which changes very quickly as the polarity is reversed on electrodes. No cataphoretic migration of particles was observed under AC current of industrial frequency but oscillations are inevitable. It may promote aggregation of crystals of new formations and partial destruction of measurable hydrosulfoaluminate films formed around cement grains at early hardening stages improving the reaction of components in the diffusion area, the slowest of all.

The effect of electric current may be felt to a greater extent in the microsstructure of concrete under the influence of temperature. Fast heating of concrete as electric current is passed through it may cause structural defects. But where the heating is gradual, this factor may not be taken into account because its influence is the same as with other methods of thermal treatment of concrete.

To sum up we can say that the cement-water system has prerequisites at the early hardening stage that its processes may be affected by an electric field. This effect may be noticeable in cement and favorable to the formation of the structure and strength characteristics of hardened cement paste.

The application of the electric field to a concrete system does not produce any pronounced effect on its structure and properties and may be neglected. Perhaps in future researchers may find parameters of and procedures for utilizing the effects of the electric field on concrete at various hardening stages such that they will produce good results which may be considered in commercial applications.

7.3 ELECTRODES FOR ELECTRIC HEATING OF CONCRETE

Electric current is supplied to concrete in electric heating through electrodes. Brought into circuit between electrodes, the concrete acts as a resistance that changes as the concrete hardens. To have proper electric heating it is necessary to choose proper electrodes and to arrange them correctly in the concrete for the heating. The electrodes should meet certain specifications. They should:

- be made of an easily available and inexpensive material;
- ensure the required area of contact with concrete to avoid its fast dehydration near the electrodes and thus the cutoff of the latter;
- minimize the consumption of metal by electrodes;
- have a clean surface;
- ensure a uniform electric and temperature field in the concrete being heated;
- preclude any chemical reaction of the electrode metal with concrete.

Thc electrodes used for electric heating of concrete may be rod, strip, band, plate, string or deposited types. Electrodes are selected as best applied to heating a specific structure according to its type, size, nature of reinforcement, and type of formwork.

Rod electrodes are made of lengths of smooth reinforcing steel bars, straight and clean of rust and oil. When electrodes are installed in concrete to be heated, their ends should protrude to a length long enough for connection to power leads. 50 to 80 mm were found to be sufficient in practice. Longer protrusions only increase the consumption of metal and do not improve electric heating in any way.

Strip electrodes are made of thin strips under one mm thick. They are placed on the surface of the formwork and are used for heating at the surface or throughout if the structure is thin.

Band and plate electrodes are made of steel strips 0.5 to 1.5 mm thick or more and are attached to the formwork or are installed on special wooden panels placed on the concrete surface. They are in contact with concrete over a larger area than strip electrodes because they may be of any width. Plate electrodes are an ideal case where the area of the electrode surface equals that of the concrete surface to be heated.

Deposited electrodes are zinc that is sprayed in strips up to 100 μmm thick on the surface of a wooden or plastic formwork. When deposited on the wooden formwork, the strips should be made along the fibers to avoid cracks in the wood, which may break an electrode. Deposited electrodes are convenient for use with permanent concrete formwork. They are cost-effective and consume 20 times less metal than required for strip electrodes that are attached to the formwork.

Steel is the most suitable metal for electrodes. Aluminum electrodes must never be used because aluminum reacts with alkalis in concrete producing a gaseous fraction that shields the electrode surface.

The area of the electrode surface in contact with concrete is of great importance. To avoid a premature cutoff of electrode because of a sharp increase of contact resistance of concrete when temperature rises too fast, the area over which an electrode contacts the concrete surface should be as large as possible. So rod electrodes should be at least 6 mm in diameter and strip electrodes at least 20 mm wide.

Correct arrangement of electrodes in concrete heating ensures their reliable operation. Certain requirements should be met there as follows:

- The distance between electrodes of different phases should be the same along the entire length, i.e., the electrodes should be installed in parallel.
- The distance of electrodes from the steel reinforcement of the structure being heated should be as great as possible and they must never contact each other. The minimum distance of an electrode from steel should be at least 25 mm.

The use and layout of electrodes are dictated by the power required to heat concrete. When plate electrodes are used for through heating and they close completely the surfaces of opposite faces of a structure, the electric output power is given by

$$P = \frac{U^2 10^{-3}}{b^2 \rho}$$

where: U = electric voltage,
b = electrode spacing in cm,
ρ = specific resistance of concrete in ohm.cm.

For practical purposes, this formula is convenient for plotting a diagram of relationships of specific output power against the electrode spacing and voltage applied when the specific output power against the electrode spacing and voltage applied when the specific resistance of the concrete is known (Figure 7.7). One can use this diagram to find output power for a specified voltage and electrode spacing. From specified electric power and voltage it is easy to determine the electrode spacing. Finally, when we know electric power and the electrode spacing, we can find the required voltage.

By substituting strip electrodes for plate ones we can reduce metal consumption considerably. Specific electric power for through heating with strip electrodes can be found from

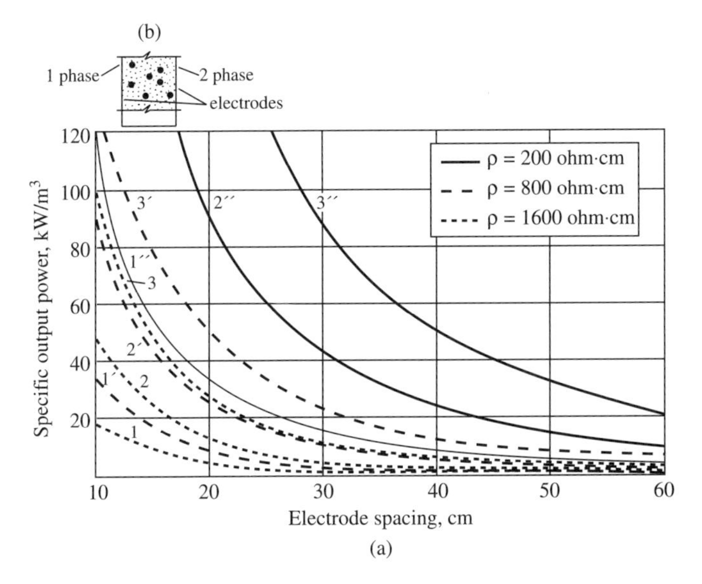

FIGURE 7.7 Specific electrical output power vs. the spacing of electrodes (a) and the layout of plate electrodes (b). 1, 1′, 1″ — voltage 51 V (ρ is 16,8 and 2 ohm/m, respectively), 2, 2′, 2″ — voltage 87 V, 3, 3′, 3″ — voltage 127 V.

$$P = \frac{U^2 10^{-3}}{\rho B^2 \left(1 + \frac{ab}{\pi B} \ln \frac{b}{2a}\right)}$$

where: B = thickness of the structure to be heated in cm,
a = width of a strip electrode in cm,
b = electrode spacing in cm.

Electric power in through heating with strip electrodes can be found in a diagram (Figure 7.7) by multiplying the obtained value by the coefficient

$$z = \frac{1}{1 + \frac{ab}{\pi B} \ln \frac{b}{2a}}$$

The values of z for different thicknesses of the structure to be heated, widths of strip electrodes, and the distance between them are given in Table 7.1.

Strip electrodes can heat concrete at the surface and can be arranged on one side of the heated structure. Electric current is passed between electrodes only on the surface layer of concrete, whose thickness is about 0.6 of the electrode spacing (Figure 7.8). This method is allowed for heating structures up to 40 cm thick with the electrode spacing being twice the thickness of a structure. When a structure is thicker than that, electrodes are installed on both sides.

TABLE 7.1
Values of z for through heating of concrete with two-sided arrangement of electrodes

	Value of z for ratio $\frac{B}{2a}$							
	0.3		0.4		0.6		0.8	
Electrode spacing	Electrode width in cm							
	2	5	2	5	2	5	2	5
10	0.944/0.925	—	0.825/0.847	—	0.754/0.704	0.952/0.934	—	0.850/0.806
20	0.862/0.825	0.971/0.971	—	0.916/0.841	—	0.800/0.752	—	0.695/0.630
30	—	0.925/0.900	—	0.854/0.818	—	0.730/0.671	—	—
40	—	0.884/0.854	—	0.820/—	—	—	—	—
60	—	0.846/0.806	—	—	—	—	—	—

Note: The numerator shows values of z for three-phase current and the denominator single-phased current.

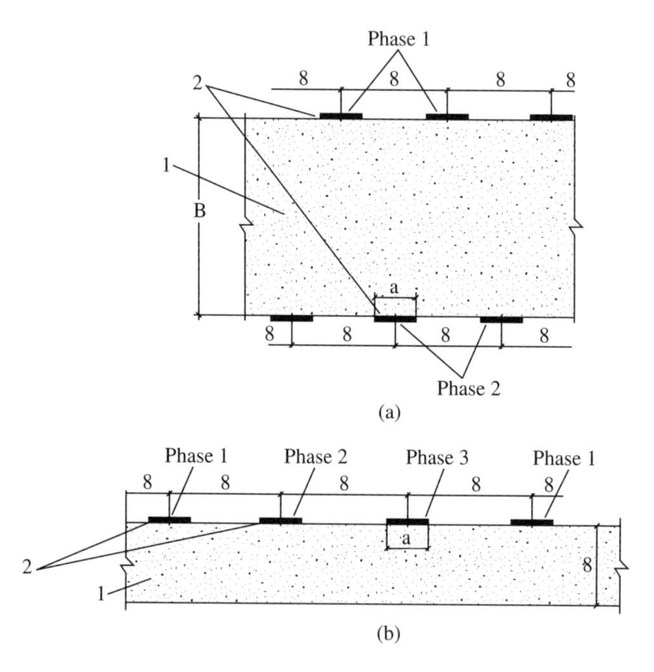

FIGURE 7.8 The layouts of strip electrodes installed on two sides and on one side for heating through the entire depth (a) and only surface layers (b). 1 — concrete, 2 — electrodes.

Specific electric power for this peripheral one-sided heating is given by

$$P = \frac{1.57U^2 10^{-3}}{\rho bB\left(a\ln\frac{4B}{\pi\alpha} + \frac{\pi b}{2B}\right)}$$

Specific electric output power in the peripheral electric heating with strip electrodes can be easily found in tables made according to this formula (see the *Appendix*). The same tables can be used to determine other electric heating parameters, such as voltage and electrode spacing on centers. It should be borne in mind that the higher is the B/b ratio the more uniform is the temperature field in the structure being heated.

Where strip or band electrodes cannot be used, rod electrodes are applied for electric heating. They are best be used not as individual rods but in the form of flat groups, each of which is connected to the same phase while adjacent groups are connected to different phases. The specific electric power in concrete heating with flat groups of rod electrodes is given by

$$P = \frac{3.14U^2 10^{-3}}{\rho bh\left(a\ln\frac{h}{\pi d} + \frac{\pi b}{h}\right)}$$

where: d = diameter of rod electrodes in a flat group in cm
h = distance between electrode in a group in cm.

The distance between electrodes in a group was assumed such that the temperature gradients in the concrete layers adjacent to the electrodes do not exceed 1°C/cm.

Nomograms of about the same type as shown in Figure 7.7 are plotted from the formula to make it more convenient to find the parameters of concrete heating with flat electrode groups.

Staggered single reinforcing bars are used in some cases, e.g., for heating heavily reinforced structures. Specific electric power for this case of electric heating can be found from

$$P = \frac{3.14U^2 10^{-3}}{\rho b^2\left(a\ln\frac{b}{\pi d} + \pi\right)}$$

For the sake of convenience in determining electric heating parameters, diagrams are plotted on the basis of this formula to find easily and quickly the required values for an equal distance between electrodes connected to the same phase or to different phases, i.e., b = h.

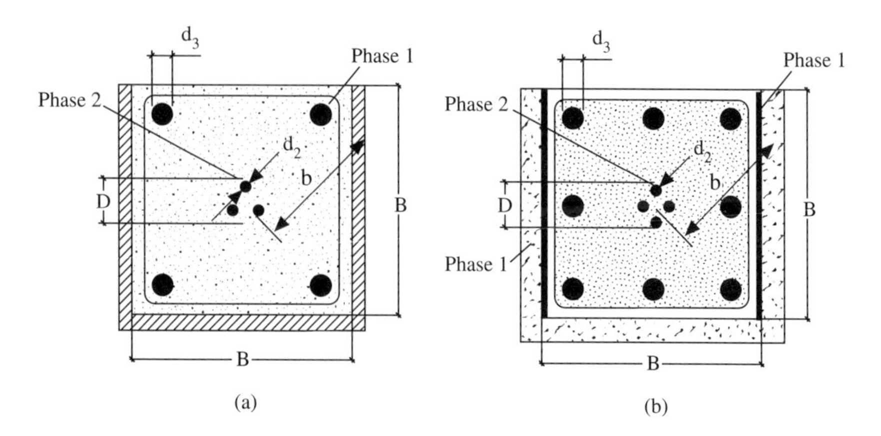

FIGURE 7.9 The layouts of wire electrodes for heating square members. a — with four reinforcing bars at the corners, b — with densely spaced reinforcing bars or in a metal form.

Single rod electrodes are hammered into concrete after it is placed and compacted to heat concrete in longitudinal joints between precast elements or thin structural members. Each electrode is connected to a different phase in this case. The specific electric power for such heating is found from

$$P = \frac{3.14U^2 10^{-3}}{\rho bB\left(a \ln\frac{B}{\pi d} + \frac{\pi b}{B}\right)}$$

where: B = thickness of the heated structure.

Strings may be either embedded into fresh concrete or installed in the middle of a structure if allowed by the reinforcement to heat concrete in the structural members whose length is many times larger than their cross section, such as beams, joists, columns, piles, etc. Reinforced concrete members of square cross sections or similar shapes are the most suitable for heating with string electrodes. If a member is to be concreted in a metal formwork or in a wooden one but lined with steel sheets (0.5 to one mm thick), a string electrode or a group of strings is installed along the centerline of the member and is connected to one phase. The formwork is connected to another phase (Figure 7.9b). The reinforcing bars of the member connected to a phase can be used in some cases as the other electrode (Figure 7.9a). Specific power for using string electrodes to heat concrete in a metal formwork is given by

$$P = \frac{6.28U^2 10^{-3}}{\rho B^2 \ln\frac{b}{d_2}}$$

where: d_2 = string diameter in cm.

If the second electrode is the reinforcement, the specific electric power can be found from

$$P = \frac{6.28U^2 10^{-3}}{\rho B^2 \ln \frac{2b}{d_2} \sqrt[4]{\frac{b}{2d_3}}}$$

where: b = distance between centerlines of strings and reinforcing bars
d_3 = diameter of reinforcing bars.

A nomogram is plotted from the formula for convenience. The required values of the heating parameters can be found there quickly and easily (Figure 7.10).

Reinforcing bars of structures should be used as electrodes only in exceptional cases, the temperature being closely watched. Uncontrolled use of this method may overdry the concrete at the interfaces with the rebars and thus reduce their bond with concrete.

It should be borne in mind in using long strip, band, or string electrodes that the voltage loss over the whole length should not exceed 5 to 8% of supplied voltage.

The choice of electrodes, their layout, and connection circuits depend on the size and configuration of a structure, the arrangement of the reinforcement, the number of structures heated at the same time, and conditions at the site.

As temperature rises in concrete, output power must not increase by more than 10 to 15% as compared with its design value to avoid local overdrying. Electric intensity during the period of heating should not be higher than 10 to 12 W/cm.

Each electric heating option has its best application and makes electric heating the most economical and efficient method for accelerating the hardening of concrete.

The peripheral electric heating is most suitable for erecting massive reinforced concrete structures. But in order to have a favorable thermal stress state in a structure, the temperature at the surface of concrete (in peripheral layers) should not rise ahead of that in the core, which increases gradually due to the heat of hydration. Rapid hardening of concrete in the peripheral layers may cause cracking due to expansion of the concrete in the core as it is heated.

One-sided peripheral heating is convenient for horizontal members with large exposed areas, such as cast-in-place concrete beds, floors, flat foundations slabs, floor slabs, bottoms of tanks, etc. up to 80 cm thick. Wooden panels with strip electrodes attached to them are installed on the surface of placed concrete. In a severe frost the panels may be covered with a heat insulator to prevent large heat loss to the environment.

It is convenient to heat such structures that have no or very little reinforcement (one mesh) with string electrodes installed in the surface layer. These electrodes are half-embedded in the concrete (they are also called *floating electrodes*) and are removed when the heating is completed.

Walls and partitions are heated with strip electrodes attached to the wooden formwork. The heating may be throughout or peripheral according to their thickness.

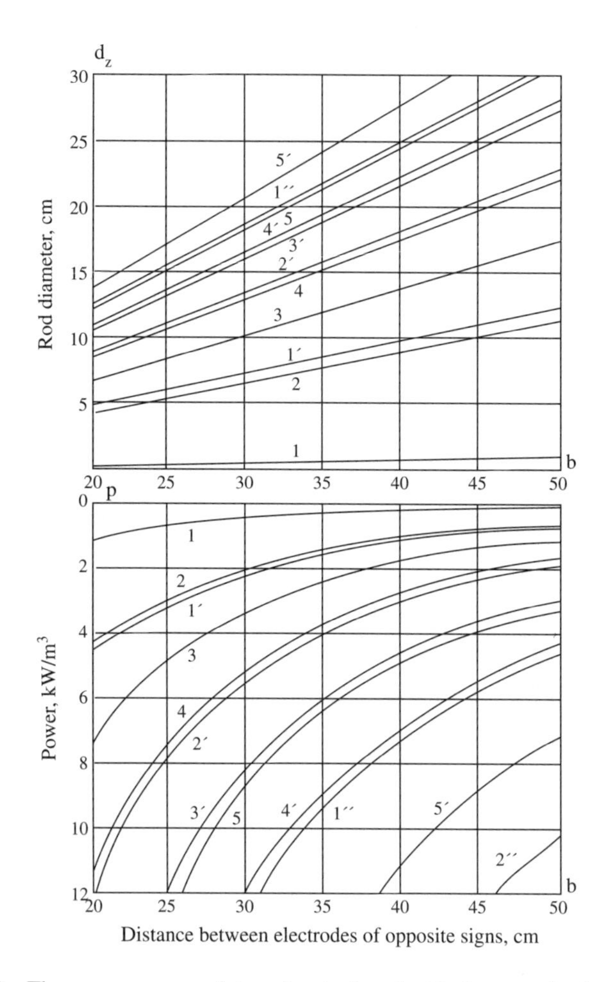

FIGURE 7.10 The nomogram used to estimate the electrical power for heating a square structural member with four reinforcing bars at the corners by wire electrodes at different voltages. 1, 1′, 1″ — 38 V (ρ is 16,8 and 2 ohm/cm, respectively); 2, 2′, 2″ — 51 V; 3, 3′ — 60 V; 4, 4′, — 70 V; 3, 3′ — 87 V

Plain-concrete or weakly reinforced foundations and beams are often heated with rod electrodes. The latter are convenient for heating concrete in joints between precast concrete components, such as columns with socket foundations, floor slabs, etc.

The capabilities of electrode heating are great and this method can be used to advantage and efficiently to accelerate the hardening of concrete in casting in place many types of structures in cold weather.

In conclusion it can be said that the correct choice of electrodes and their proper arrangement can ensure electric heating of concrete in keeping with designed conditions until the concrete attains the required strength.

7.4 PROCESSES AT THE INTERFACE OF CONCRETE AND AN ELECTRODE

In electric heating of concrete, we encounter a number of electrochemical, electro-physical, and simply physical phenomena where it contacts an electrode. Insufficient knowledge of these phenomena gave rise to the wrong opinion that concrete always overdries in the vicinity of electrodes and, moreover, to a sceptic attitude to this promising and efficient method for the acceleration of concrete hardening in general. Both are fallacies that can be dispelled by a careful analysis of the phenomena that occur in the electrode area.

Studies conducted by A. K. Reti, V. P. Ganin, and V. Ya. Gendin demonstrated that a transitional resistance appears at the interface of an electrode with concrete. It can vary over a wide range according to various influence factors. But phenomena other than the transitional resistance occur and their impact on electric heating is predominant.

The NIIZhB has carried out specific investigations of electrode phenomena and their findings made it possible to heat concrete electrically more competently and efficiently.

As was mentioned above, electrochemical, electrophysical and physical phenomena occur at the interface as electric current passes from electrodes into concrete during electric heating.

The electrochemical phenomena are associated with the resistance that arises at the interface of materials of different conductivities. The contact resistance appears in electric heating at the interface of ion and electron conductors as a result of the transition from the electron (metallic electrode) to ion (liquid phase with dissolved electrolytes) conductivity. Polarization may be an obstacle for electric current in this case although its influence is negligible on electrodes of a small area and at a frequency of more than 20 Hz. When concrete is brought into circuit, a double electric layer is formed on the electrode and it is also an obstacle to the current passing from metal to concrete. This layer is formed by electric charges on the metal and by ions of opposite sign in the concrete's liquid phase at the surface of the electrode. Rise in temperature accelerates migration of electrolyte ions dissolved in the liquid phase toward the electrode surface, which should improve the current conducting properties in the boundary layer. Without dwelling on an analysis of the formation of the double layer and of the factors that affect it we just note on the basis of our findings that the electrochemical phenomena, although they do occur, are not very important to electric heating of concrete and can be neglected.

A much greater impact on the contact resistance is produced by electrophysical and physical processes that go on in concrete in hardening. While the liquid phase of concrete, which is a homogeneous system, is in close contact with the electrode surface, the concrete mixture is a heterogeneous system and cannot have such a contact with the electrode. A part of the interface area is covered by bubbles of air entrapped in the concrete and by voids left in vicinity of electrodes due to faulty vibration of the concrete, particularly when stiff mixtures are used.

Experiments disclosed that even after a good compaction by vibration air bubbles still cover up to 2% of the contact area at the interface of an electrode with concrete. When the mixture is stiff and vibration bad, the contact area covered by air bubbles may increase many times. At normal temperature, the reduction of conductive cross section at the interface by several percent does not cause any sharp rise in resistance in this cross section.

But the picture is quite different when the temperature rises, particularly with accelerated heating of concrete mixture. It is accompanied by an increase in the volume of gas bubbles whose cubic expansion coefficient is higher than that of the solid phase by several orders of magnitude and that of the liquid components of concrete by an order of magnitude. Moisture starts to evaporate in the cavities and the bubbles are filled with water vapor.

The expansion of the vapor-air mixture increases with rise in temperature. This increases internal pressure that expands the air voids at the interface and thus pushes a part of the concrete away from metal. Moreover, the pressure may exceed the force of adhesion of the concrete to the electrode and form channels for migration of the gaseous phase to the surface of the structure.

Heating of concrete with thin rod electrodes or with narrow strip electrodes is associated with the appearance of a zone of higher current density at the interface. Its temperature rises fast and may reach 100°C if the design and layout of electrodes were wrong or the heating was not closely supervised. The liquid phase starts to boil and the gas bubbles released at the contact surface go up and come outside. This considerably disrupts the close contact of the concrete with the electrode.

Finally, the quality of the electrode surface is quite important. Experiments and observations at the site proved that the contact of concrete with an electrode may be impaired for three reasons, including the presence of hardened concrete particles on the electrode surface, uneven or excessive lubrication with nonconducting substances, and the appearance of oxides on a steel electrode. Several of these factors often act together.

Hardened concrete particles that remained on the electrodes due to poor cleaning can remove large areas on the contact surface of the material from direct heating. It is not only the high resistance of hardened or stuck particles that presents an obstacle to current but also the great number of air bubbles caught by the rough surface of the electrode.

Lubricating oil, usually dielectric, unevenly applied to the electrode surface also isolates some parts of the concrete surface from heating or considerably impairs the flow of current from the electrode to the concrete.

Oxide films that form on the electrode surfaces have a high resistance and thus shield some areas. Diffusion may occur in the oxide films only at high temperatures. Characteristically, metal diffuses to the outer surface of a film from the metal-film boundary in the form of ions together with migration of electrons. The electrons join oxygen at the outer surface forming new oxygen ions. The oxide film usually has uneven thickness thus making the surface rough. This hinders the uniform flow of current into concrete and does not let air bubbles stop at the interface.

The gaseous phase expands with rise in temperature, as was mentioned above, and starts to migrate to open surface of concrete along the electrode leaving distinct

FIGURE 7.11 The appearance of the faces of concrete specimens, which contacted a clean polished electrode (left) and an uncleaned electrode with traces of corrosion (right).

migration traces. A comparison of contact faces of concrete that touched a clean polished electrode and the one that was cleaned of rust presents a striking contrast (Figure 7.11).

An analysis has shown that the contact fault due to the appearance of the gaseous phase (F_1) on the electrode can reduce the effective area of a flat electrode by 20 to 50%, which immediately increases the current density and raises the temperature in this zone.

The considerable shielding of an electrode can be avoided at the site by using plastic concrete mixtures; good compaction of concrete, particularly near the electrodes, and by using electrodes with clean surfaces.

The current conducting cross section of the concrete is thus reduced at the interface by F_1. The drop in F_1 increases many times when concrete is not dense or has large pores.

The area of close contact of concrete with the electrode is far from uniform in its conducting properties. A part of it (F_2) is occupied by coarse aggregate and part (F_3) by fine aggregate. Naturally, these values may vary according to the type and composition of concrete, grading of the aggregates, and their type.

As was pointed out above, aggregates have a much higher specific resistance than hardened cement paste and so can be safely called nonconducting inclusions. Granules of coarse and fine aggregate contact electrodes through a thin film of cement paste. It does not reduce the conducting cross section of concrete at the interface with an electrode at the early heating stage but later on decreases it as the film dehydrates.

Cement grains, though finely dispersed particles, are incomparably larger than molecules or ions. So they should have been called dielectric inclusions too. But, considering the presence of physically bound water films around cement grains, where the process of dissolution of cement clinker minerals continues, and considering that the cement grains contact the electrode through this current conducting film, we can roughly consider the cement paste a homogeneous conducting system. This assumption is acceptable only for the early stage of concrete hardening when dense shells of crystals of new formations have not yet appeared around cement grains.

So the cross section of the current conducting part of concrete mixture at the interface with an electrode is $S = F - \Sigma F_n$, where F is the contact area of the electrode with concrete and ΣFn is the sum of the areas of a contact face, which are shielded by the gaseous phase and are occupied by dielectric inclusions ($F_n = F_1 + F_2 + F_3$).

The dehydration of concrete near electrodes, which sharply increases the contact resistance, may happen also for another reason — wrong layout of electrodes, their small surfaces, and poor contact with concrete. The reason for the dehydration in the first and second case is an excessively high current density at the interface of the electrode with the concrete, which overdries the latter. The conducting cross section is reduced in the third case and so the amount of released heat increases according to Joule's law. This quickly overheats and overdries conducting bridges. Concrete is overdried not only because moisture evaporates from it to the environment (at the surface of the structure) but also because the moisture transfer potential appears as the thermodynamic equilibrium in the system is upset. Rise in temperature at the interface results in migration of the flow of moisture from the area near the electrodes down to colder layers of the concrete.

When concrete is heated in the condition where its exposed surface is covered with a vapor and heat insulator, the dehydration of the electrode area results in a gradient of the moisture content, which is opposite to the temperature gradient. So forces appear, which try to transfer moisture from the core to the electrode area. Since moisture conduction forces and thermal moisture conduction forces have different directions, they soon come into equilibrium.

A high thermal gradient arises in electric heating of structures with an exposed surface between the concrete in the area near the electrodes and the environment, which causes moisture to evaporate quickly and dehydrate the concrete.

The above-described factors that affect contact resistance all relate to the case when an electrode has a close contact with concrete. However, in electric heating with electrodes attached to the formwork they may be separated from concrete and thus an air gap may appear. The separation is caused by shrinkage of concrete, particularly at the later hardening stage. The air gap impairs the contact of the electrode with the concrete and thus makes further heating impossible because the air becomes an insulator even in a layer measured in fractions of a millimeter.

Sometimes, if concrete had not yet attained the required strength by the time when the electrode was separated, the heating may be prolonged by introducing a weak electrolyte solution in the gap. To reduce the viscosity of the solution and for better filling of the gap, the solution should be warmed up to 50 or 60°C. The electric heating is best to be continued after that at a temperature not higher than 50°C to avoid fast evaporation of the electrolyte. The concrete surface in the contact layer must be covered with a vapor and heat insulator.

This brief discussion of the kinetics of the processes taking place at the electrodes in the course of electric heating of concrete shows their great importance. Contact resistance, which is the consequence of various actions may grow quite high and become an obstacle in electric heating. Precautions in practice should meet the following requirements:

- The electrode surface should be clean and smooth, without oxides or particles of hardened cement paste stuck to it.
- Concrete in the area near the electrodes should be compacted with a particular care.
- Electric heating should not be too fast to avoid excessive thermal gradients.
- The exposed surface of the structures to be heated should be covered with a vapor insulator and a heat insulator to avoid loss of moisture to the environment.

These precautions allow to carry out the electric heating of concrete for a long time and to produce structures of high quality with low consumption of electric power.

7.5 ELECTRIC AND THERMAL FIELDS IN ELECTRIC HEATING

Uniformity of the electric and thus thermal field in electric heating of concrete is of a particular importance because the method has an excellent ability to raise temperature across the whole cross section of a structure within any time period required. The uniformity of the heating depends on the way the electric field is formed. It is greatly affected by the reinforcement because of the great difference in electric conductivities of steel and concrete (it is 10^8 higher in steel).

The reinforcement usually adversely affects the formation of the electric and thermal fields and may be neutral at best, which rarely happens.

Under mains voltages used for electric heating of reinforced concrete structures the layout of the reinforcement has a profound effect on the fields. Starting power may increase considerably, the concrete near steel may overheat, moisture may evaporate quickly, and the concrete may be dehydrated and overdried in some parts of a structure.

It should be stressed that to ensure the uniformity of heating reinforced concrete structure is a very difficult task and few researchers undertook to study this problem. Partial solutions for different layouts of electrodes were proposed by A. V. Netushil and R. V. Vegener but they did not provide the whole picture of the formation of electric and thermal fields in structures. The NIIZhB has conducted extensive in-depth studies of this problem. They used physical models that were geometrically similar to the cross sections of reinforced concrete elements. This model was used to demonstrate the distribution of equipotential and current lines for scores of variants of reinforcement arrangements and the way they changed with changes in the layout of electrodes.

An analysis of the data thus produced revealed that when the reinforcement was arranged at right angles to the direction of electric current, it did not affect the uniformity of the electric field for solid plate and strip electrodes (Figure 7.12). The picture was no different even when several disjoint meshes were installed. The intensity of the field in this case can be assumed to be uniform and can be calculated using the standard formula

$$e = \frac{U}{h}$$

where: e = field intensity in W/cm,
U = voltage supplied to electrodes in V,
h = thickness of a structural element being between electrodes in cm.

Where several parallel meshes were placed between electrodes at right angles to current, the distribution of equipotential lines remained uniform. But metallic inclusions that took up a part of the volume of the heated body reduced the total resistance between the electrodes and increased the strength of the current (Table 7.2). The strength did not change when the reinforcing mesh at right angle to the current was positioned at different distances from the electrodes.

For practical purposes, the reinforcement in the form of one or two floating meshes may not be taken into account because the volume of metallic inclusions occupies no more than 3 to 6% of the cross sectional area and the strength of current changes only by a few percent, i.e., only slightly.

So in electric heating of structures reinforced with disjoint bars or meshes arranged at right angles to the direction of current the uniformity of the electric field does not vary and thus there are no noticeable changes in the thermal field too. Only boundary conditions may affect the uniformity of the latter in structures with this type of reinforcement.

The situation is quite different when the reinforcing elements between the electrodes coincide with the direction of electric current. The shielding effect of the cage distorts the electric field in this case. The current passes mainly through the concrete cover and the thinner the latter the higher the intensity of the field in it. The cover thus heats up most of all and a fast rise in temperature may overdry the concrete. The heating may even stop as the current cuts itself off due to highly increased resistance in the dehydrated area. So structures reinforced with cages can be heated with electric current only gradually so that heat released in the concrete cover has time to relax to the core due to heat conduction and convective heat transfer to the environment. When the cover is less than two to 2.5 cm thick in heavily reinforced elements, it is very difficult to heat them by the electrode method even under very gradual heating conditions.

When a structure is reinforced with simple, not dense cages (usually four longitudinal bars tied together with stirrups), the electric field is not uniform to a great extent in the plane of the stirrups. It levels off gradually between the stirrups and

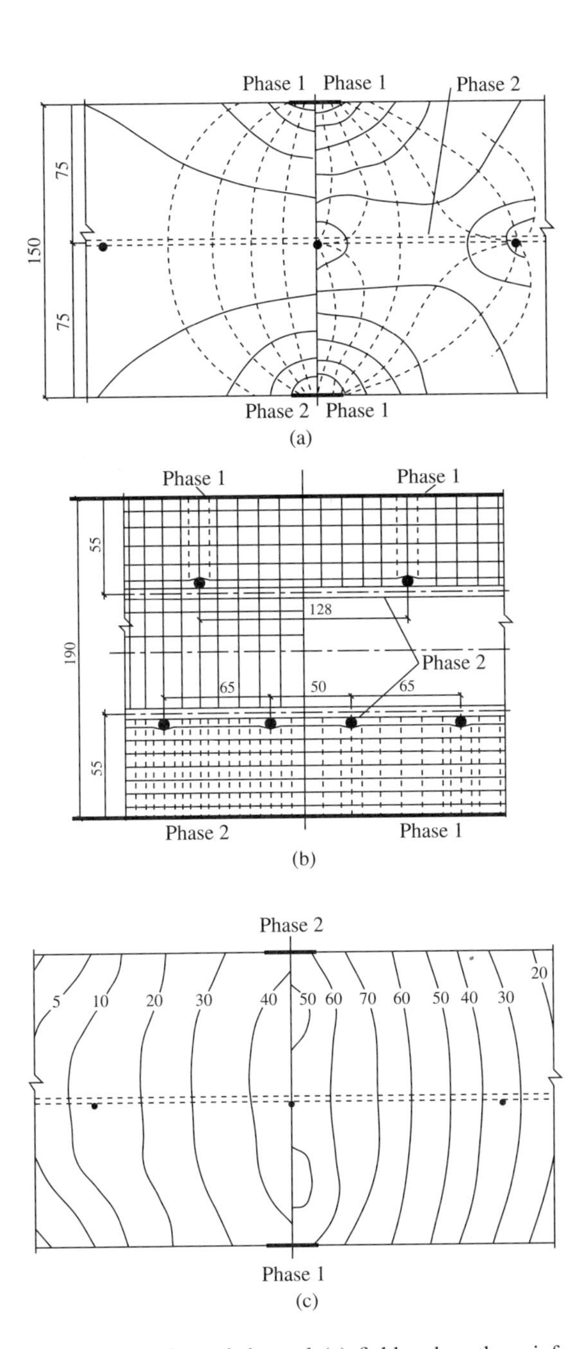

FIGURE 7.12 Electric (a and b) and thermal (c) fields when the reinforcing meshes are located at right angles to the direction of current. Figure 7.12c shows a thermal field after one hour of heating on the left and that after four hours on the right.

TABLE 7.2
Variation of current strength according to the amount of reinforcement in the form of meshes placed at right angles to the direction of the current.

Strength of current in mA	Volume of reinforcement in mL				
	0	1	2	12	24
I	26.5	26.8	27	28	29.2
$\frac{I_i}{I_0}100$	100	101	102	105	110
$\frac{I_i^2}{I_0^2}100$	100	102	104	112	122

where: I_o = current strength in electrolyte without reinforcement
I_i = current strength in electrolyte with reinforcement.

becomes relatively uniform at a distance equal to about two thicknesses of the cover from them and the concrete is heated over the whole cross section.

Full shielding of internal layers of concrete in a structure occurs only with the reinforcement made of rolled sections (e.g., I-bars with their flanges facing the electrodes), with very dense reinforcing cages, or when electrodes are placed along the longitudinal reinforcement (grounded or connected to the phase) in front of bars (Figure 7.13).

With simple reinforcing cages, the formation of the thermal field corresponds to the distribution of the current lines. When electrodes are placed in the plane of stirrups, the temperatures are not uniform to a great extent from the very beginning of heating, the difference being as high as 50°C within the stirrup spacing (Figure 7.14). Heat is released mostly in the concrete cover between the electrodes and the stirrups. Characteristically, after four hours the temperature difference reaches 70°C due to the temperature rise in this region, which threatens to overdry the concrete if special precautions are not taken.

When electrodes are placed in the plane of stirrups, the concrete is heated inside the cage solely due to heat conduction from warmer parts and heat of hydration. This area is beyond reach of the electric current.

The thermal field is formed in a somewhat different manner when electrodes are located between stirrups (Figure 7.15). The temperature difference in this case does not exceed 25°C at the start of heating and rises to 55 or 60°C within four hours. Although heat is released mainly in concrete covers, some rise in temperature also takes place between the stirrups. It is no longer no man's land and lies within reach of the electric current.

The connection of the reinforcing cage to the phase complicates the electric and thermal fields. Then concrete is warmed up by Joule's heat only in the concrete covers and there is practically no electric heating inside the cage. Rise in temperature

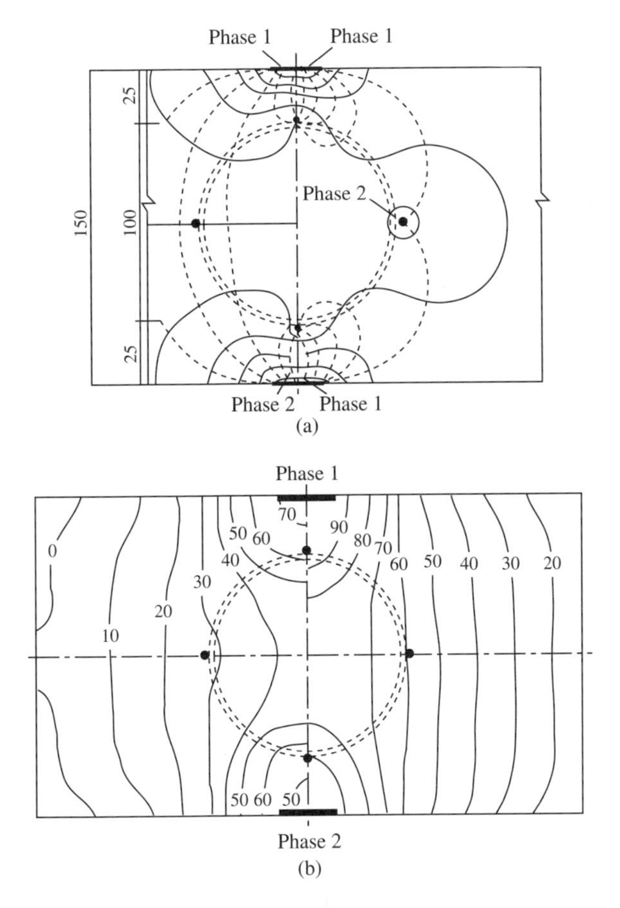

FIGURE 7.13 Electric (a) and thermal (b) fields when the electrodes are located along the reinforcing cage in front of longitudinal bars. Figure 7.13b shows a thermal field after one hour of heating on the left and that after four hours on the right.

in the interior areas that are beyond reach of electric current is caused by heat conduction and heat of hydration (Figure 7.13).

Electric and thermal fields also agree quite well when electrodes are placed on one side of a structure, which is typical of peripheral electric heating.

With this heating arrangement, noteworthy is the case when an electrode is installed in front of a stirrup. Here the electric field becomes extremely nonuniform and the current density in the cover is especially high in the region between the electrode and the stirrup making this area particularly dangerous as far as overheating is concerned (Figure 7.16).

When an electrode is placed between stirrups, the electric field becomes more uniform (Figure 7.16). Despite the difference between the electric field produced by electrodes located in front of stirrups and the field that arises when the electrodes are placed in the interval between them, there is no much difference in the thermal fields as early as after four hours of heating. The reason must be a more intensive

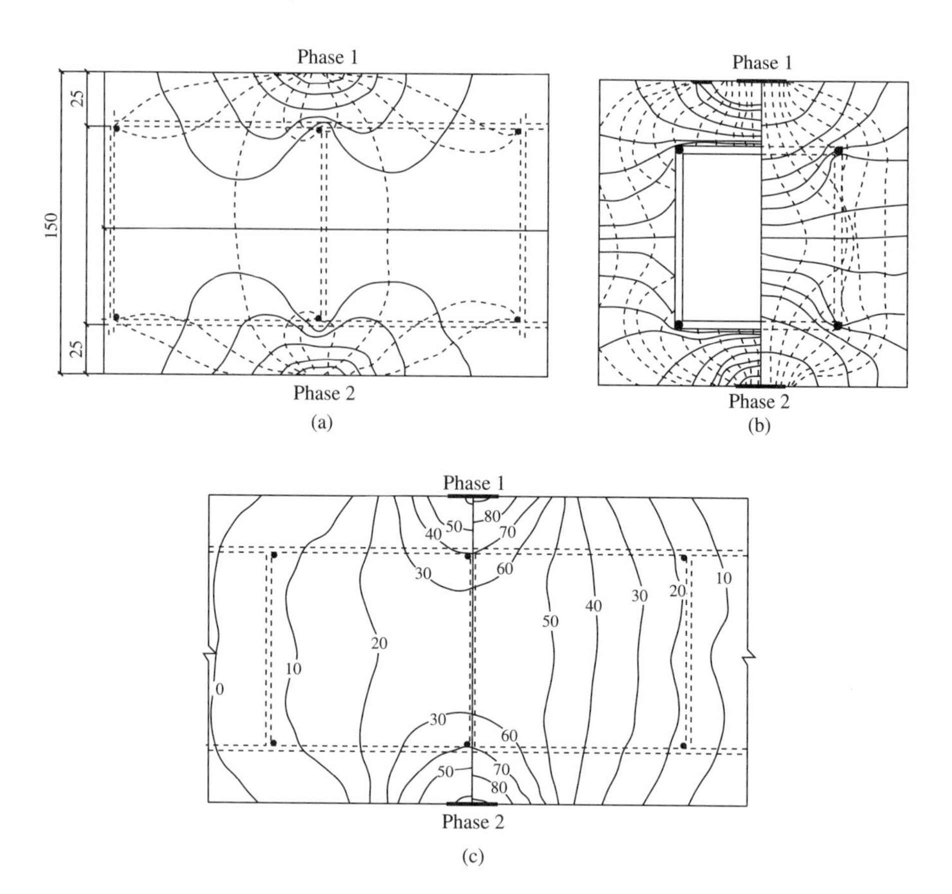

FIGURE 7.14 Electric (a and b) and thermal (c) fields when the electrodes are located in front of the reinforcement stirrups. Figure 7.14b shows an electric field in the plane of a stirrup on the left and that between stirrups on the right. Figure 7.14c shows a thermal field after one hour of heating on the left and that after four hours on the right.

heat transfer at high thermal gradients that arise across the cross section of the element being heated. Heat of hydration undoubtedly affects the temperature in the cover of concrete.

The foregoing gives reasons to believe that with a sufficient thickness of the cover and gradual electric heating the cage-type reinforcement affects considerably the electric and thermal field only at the initial heating stage and much less at the final stage.

Floating bars connected to the phase are used sometimes in electric heating of reinforced concrete structures. A study of electric fields in this type of heating disclosed that they were extremely nonuniform with the predominant concentration of current lines in the area of the floating bar acted as an electrode (Figure 7.17). For this reason, floating electrodes are seldom used although their use is feasible when heating is very gradual. The diameter of the bars should be as large as possible but not less than 10 mm.

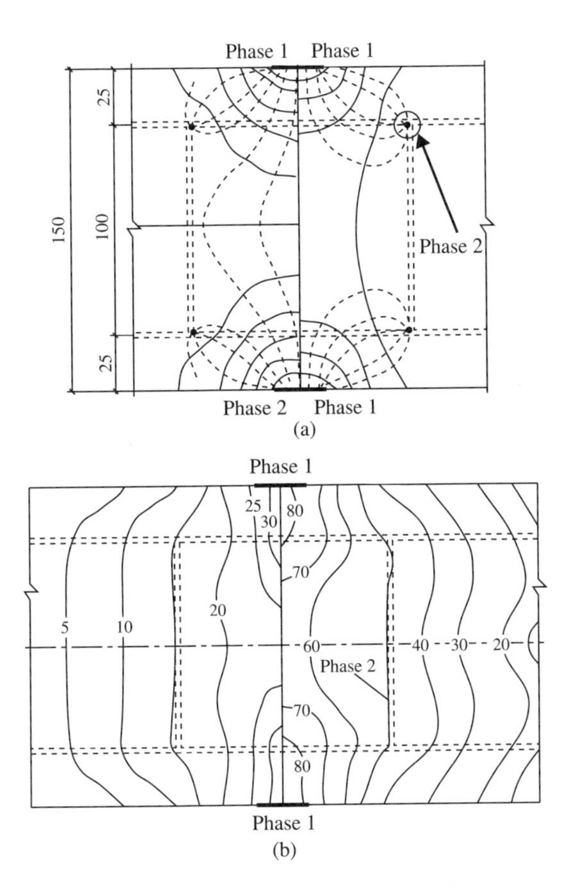

FIGURE 7.15 Electric (a) and thermal (c) fields when the electrodes are located between the reinforcement stirrups. Figure 7.15b shows a thermal field after one hour of heating on the left and that after four hours on the right.

The findings of experiments with electric and thermal fields have shown the existence of certain regularities. However, it was impossible to establish a strict analytical relationship between the fields and the way they are formed. This problem is very complicated and the mathematical description of these phenomena for a three-dimensional system that, moreover, relaxes with time is very difficult. The research in this area should be continued. An analytical solution understandable for construction engineers is sure to be found in future to make it possible to predict the formation of electric and thermal fields in electric heating of all types of reinforced concrete structures.

7.6 EXTERNAL MASS EXCHANGE IN ELECTRIC HEATING

The presence of temperature and moisture gradients between concrete and the environment causes the phenomenon of mass exchange. It becomes particularly

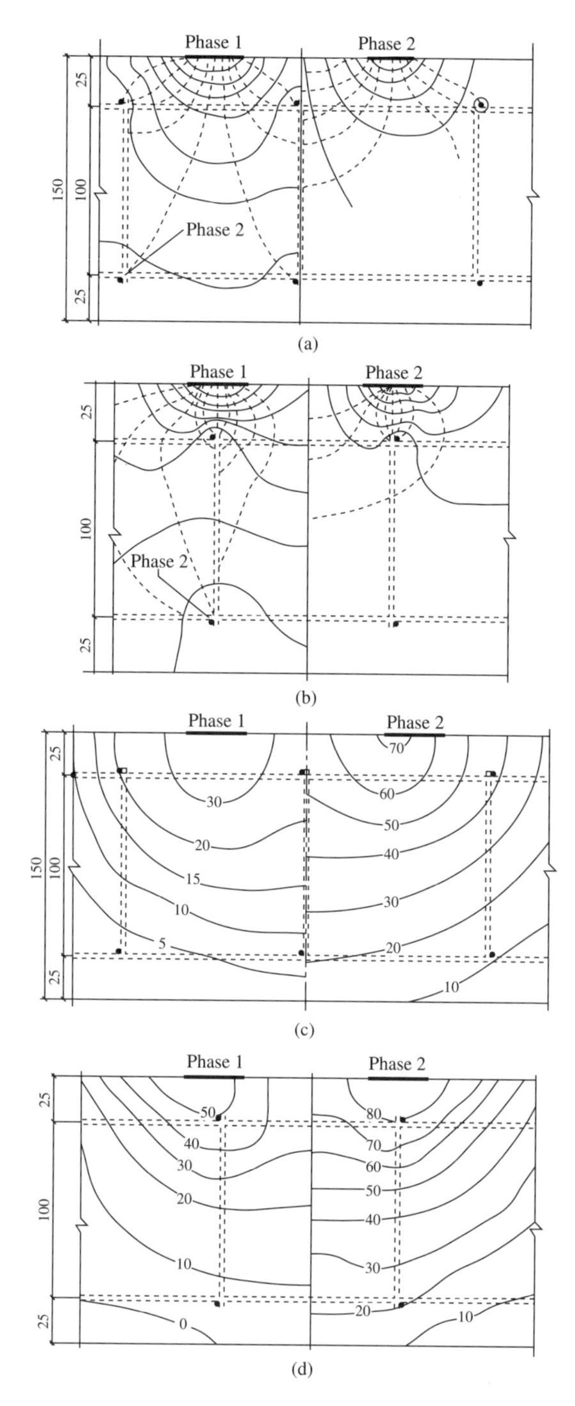

FIGURE 7.16 Electric (a and b) and thermal (c and d) fields when the electrodes are located on one side in front of the reinforcement stirrups and between them. Figures 7.15b and 7.15d show thermal fields after one hour of heating on the left and that after four hours on the right.

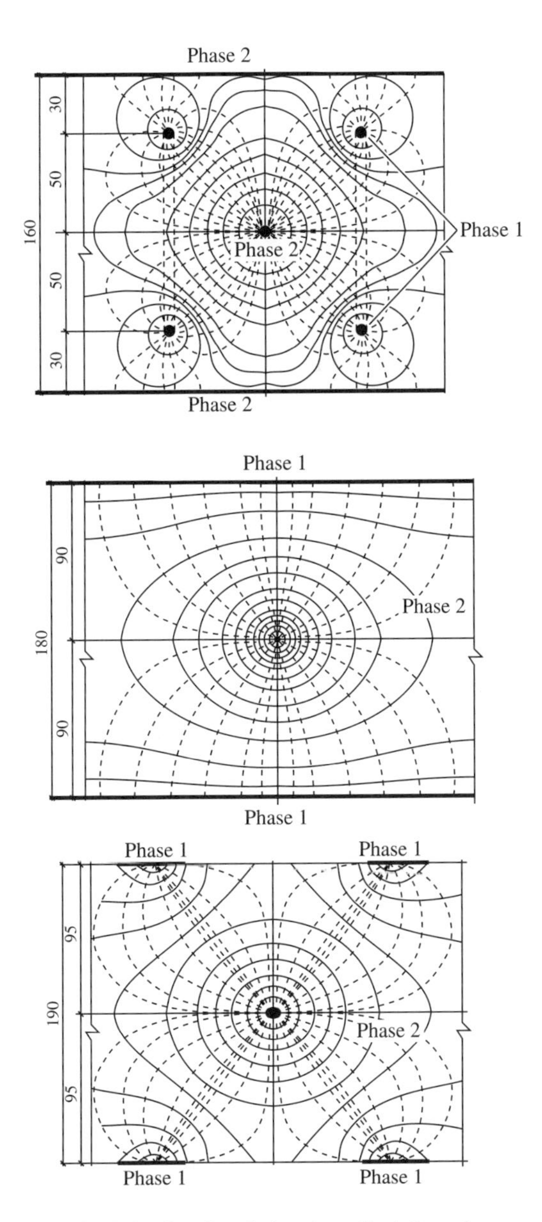

FIGURE 7.17 Electric fields in the electric heating of reinforced concrete structures using a floating rod connected to the phase.

important in electric heating of concrete (electrode heating) since the supply of heat with this method of thermal treatment is quite different from the heating procedures whereby heat is supplied to the surface of concrete. In electrode heating, rise in temperature takes place across the whole cross section and the temperature gradient between it and the environment remains during the entire heating period. The flow of moisture in this case is directed from concrete to the environment at all heating stages. Although an excessive quality of water is added to concrete in mixing, the material may dehydrate and its chemical reactions may be retarded as the water rapidly evaporates.

However, more than 60 years of application of the electrode heating in practice have shown that the dehydration and overdrying of concrete with this method of thermal treatment happens only when the heating conditions were selected incorrectly and there were errors in the layout of electrodes. At the same time, moisture loss may be a good thing for the quality of concrete if it does not exceed admissible values and does not cause structural defects. Reasonable dehydration may increase its density (if concrete has not lost yet its plasticity) and improve thermal physical characteristics, which is particularly important for structures made of lightweight concrete or cellular concrete.

Many Russian experts have studied the problems of mass exchange in concrete but their investigations were mainly occasional, which made any generalization quite difficult. Extensive studies were conducted at the NIIZhB under my guidance on mass exchange to determine primarily the effects of procedures and environmental conditions on the kinetics of the moisture loss by concrete in the process of heating and then curing.

Moisture is lost to the environment when there are temperature and moisture gradients that disturb the equilibrium moisture condition of a system. Trying to restore the disturbed state of equilibrium, the capillary-pore system (which is concrete among others) starts to give off its moisture. We can reach the equilibrium state again either by desorption or by increasing the humidity and temperature of the environment. Thus the hydrothermal equilibrium becomes possible, wherein vapor pressure in the material is equal to the partial pressure of vapor and air, and consequently the evaporation process stops (this statement holds at $t = \text{const}$).

Although mass exchange and heat exchange with the environment take place from the surface of the material, they are closely associated with moisture and heat transfer inside it. The mechanism of the mass and heat transfer inside concrete is very complex and depends on the way moisture is bound at each hardening stage. But still the kinetics of moisture loss at the start, i.e., at the early hardening stage, is governed by external heat and mass exchange.

In electric heating, a peculiarity of the moisture-content field is that the quantity of water in the material diminishes from layer to layer upwards to the evaporation surface. The core lags behind in the subsidence of moisture while the surface layers lose it quickly. According to Prof. A. V. Lykov, the moisture transfer into the material mostly has the form of vapor at high temperature gradients. At the early heating stage there is an internal moisture transfer along with starting evaporation. It increases with rise in temperature that promotes the flow of moisture to the boundary surface.

When the temperature of the material exceeds 60°C, a gradient of total pressure arises inside it, which is favored by slippage diffusion in microcappilaries and by effusion inflow of air in microcapillaries (the temperature in the core is higher than on the evaporation surface). Characteristically, in electric heating the temperature gradient not only does not hinder the migration of moisture to the surface (as it happens, for example during initial heating and isothermal curing in steaming) but favors it because their directions coincide. True enough, the concentration diffusion of moisture directed to the core from the concrete surface (the moisture content on the surface is higher than in the core at the early stage) puts up some resistance to the moisture transfer to the surface but its effect is reduced to zero as the concrete heats up.

Evaporation of water in electric heating takes place in the whole volume of concrete, more so in the core on the surface because of higher temperature there. This is the reason for the appearance of the gradient of total pressure, which is the main moving force of vapor transfer inside the material.

The foregoing explains well the occasional explosions of concrete caused by local overheatings inside when electrodes were arranged incorrectly. I have seen it happen when temperature was increased rapidly in electric heating of precast concrete elements with mesh electrodes. Since internal electrode meshes sometimes had irregular openings due to poor fabrication quality, the temperature on an electrode rose to 100°C. Internal evaporation reached dangerous proportions because there was no outflow of vapor due to high density of the material. The pressure in vapor pockets reached critical values, which broke off large lumps of concrete that weiged sometimes several kilograms.

The nature of internal evaporation was clearly demonstrated when experimental specimens of normal weight concrete were heated rapidly. Internal cavities produced by the pressure of vapor due to evaporation of water inside the material were noticeable visually in the specimens where the rate of the rise in temperature exceeded 20°C/hr. When the temperature rise was 80°C/hr, the number and size of the cavities increased sharply, the surfaces of the specimens heaved and radial cracks appeared on it (Figure 7.18).

The applied electric field in electric heating also produces moisture transfer under the action of thermodynamic force. But experimental data on electric diffusion of water are not available and it is impossible yet to estimate the importance of this factor in the total moisture flow.

Experiments with the moisture loss from different concretes in heating found that the evaporation from normal weight concrete and from lightweight concrete with expanded clay proceeds in the same way and the difference lies only in the amount of evaporated moisture. Various factors affect the loss of moisture from concrete in heating.

The effect of water content on the moisture loss from concrete. The initial amount of water added to concrete in mixing does not affect the nature of evaporation from the exposed concrete surface but affects its quantity (Figure 7.19). Other conditions being equal, concretes containing less water lose higher percentage of its initial content although in absolute values the moisture loss is higher in concretes that had more water at the start. It can be seen under different heating conditions (Table 7.3).

The higher absolute moisture evaporation from concrete with a higher water content is the result of a change in its physical moisture properties. It is common knowledge that the porosity of concrete increases as the water content rises. Inflow of moisture from the core to the evaporation surface is faster and larger while the moisture loss into the air increases when the surface is larger (water evaporates also from pores open on the surface).

FIGURE 7.18 Structural deterioration of concrete in specimens heated at different rates.

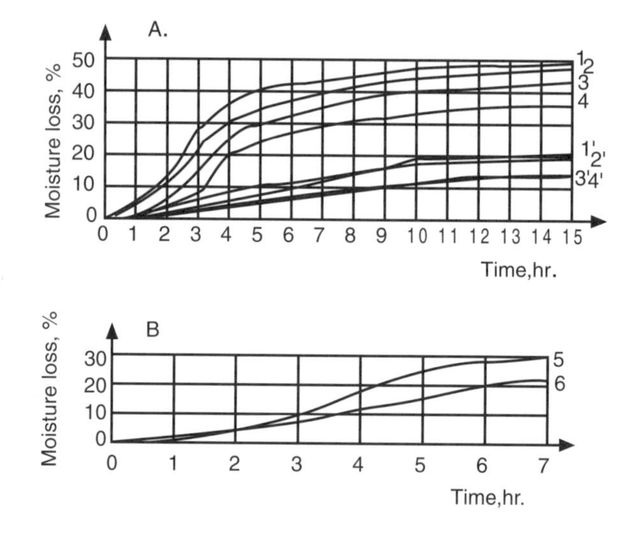

FIGURE 7.19 Moisture loss by concretes in the process of electric heating at 80°C according to the water content. a — normal weight concrete, b — lightweight concrete, 1 to 6 — exposed specimens, 1′ to 4′ — covered specimens. Water content: 1, 1′ — 135 l/m^3; 2, 2′ — 195 l/m^3; 3, 3′ — 225 l/m^3; 4, 4′ — 325 l/m^3; 5 — 190 l/m^3; 6 — 290 l/m^3.

TABLE 7.3
The amount of evaporated moisture from concretes with different water contents and heated at different temperatures

Type of concrete	Heating time in hrs.	Heating temperature in °C	Water content per 1 m³ of concrete in l	Water content of specimens in grams		Evaporation of moisture from specimens in grams	
				not covered	covered	not covered	covered
Normal weight concrete	3+8+4	60	135	227	213	93	27
			195	368	360	150	26
			225	423	407	155	35
			325	597	587	172	52
		80	135	217	213	109	45
			195	367	364	178	73
			225	425	409	185	55
			325	590	586	219	77
	3+12+4	60	135	215	214	93	35
			195	366	370	147	32
			225	416	421	152	31
			325	594	590	188	46
		80	135	216	214	100	51
			195	370	373	159	71
			225	416	423	164	80
			325	575	588	247	117
Lightweight concrete	1+3	80	190	346	349	63	10
			290	561	564	75	15
		98	290	585	524	127	15
	3+3	80	190	352	358	97	17
			290	586	587	115	24
		98	290	527	553	144	12

Notes: The heating times are given for normal weight concrete at three stages, including the time of rise in temperature, isothermal curing time at the given temperature and cooling time; and for lightweight concrete with expanded clay at two stages, including heating time and isothermal curing time at the given temperature.

Studies demonstrated the need to cover the exposed surface of structures being heated at all times to protect it against the moisture loss, particularly if the concrete mixture contained little water.

The effects of the exposed-surface modulus on evaporation of water. The great variety of plain and reinforced concrete structures makes us give special attention to the presence of exposed surfaces (without formwork) from which heat and moisture are lost in thermal treatment. The exposed-surface modulus is a ratio of the area of the exposed surface of a structure to its volume. The amount of evaporated moisture in exposed specimens was found to increase with an increase in the exposed-surface modulus linearly at any heating temperature. The loss of moisture is particularly dangerous for thin structural members. Thus, specimens

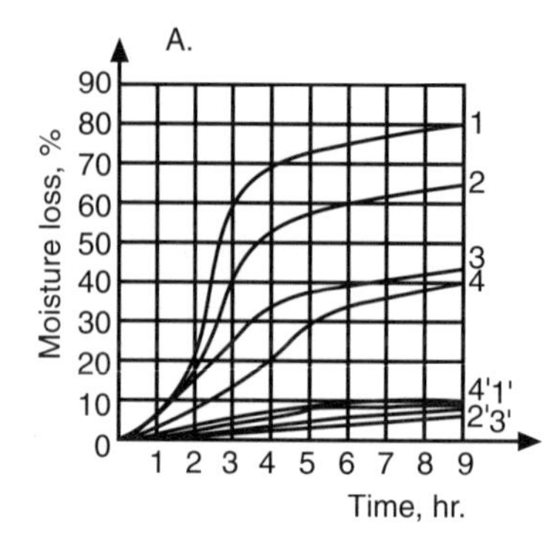

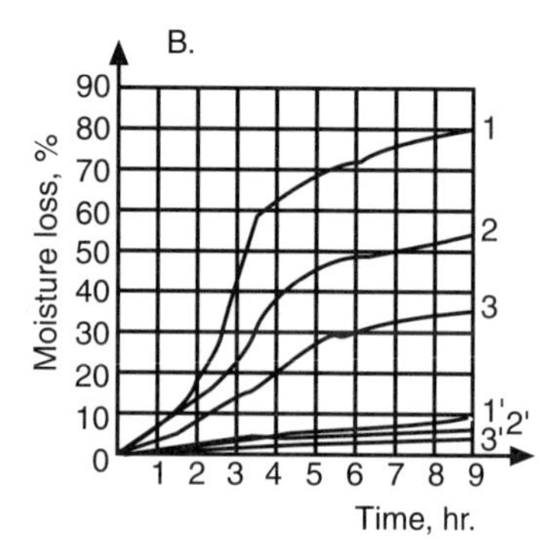

FIGURE 7.20 Evaporation of moisture from concrete specimens with different moduli of the exposed surface. a — heating at t = 80°C and b — heating at t = 60°C. 1 to 4 — exposed specimens; 1′ to 4′ — covered specimens, 1, 1′ — specimens with M = 40; 2, 2′ with M = 20; 3, 3′ — with M = 10; 4, 4′ with M = 5.

2.5 cm thick (the modulus of the exposed surface was 40 with the dimensions being 10 × 20 cm) lost up to 80% of water after nine hours from the start of heating (Figure 7.20).

A lower isothermal curing temperature of 60°C practically did not change the situation and the amount of evaporated moisture still was 77°C, i.e., decreased by only 3%. Heavy loss of moisture instead of admissible 12 or 15% damages the concrete and greatly retards further development of its strength.

The total moisture loss dropped dramatically in thicker specimens. But it was different in layers that were at different distances from the evaporation surface. The top layer whose thickness was equal to the depth of a 2.5-cm specimen lost the same amount of moisture as the previous specimen. The moisture loss higher than at a later heating stage took place due to migration of water from the core. The process was retarded somewhat because the evaporation area went deeper. Moisture moved up to that area from the interior as a liquid. But it went to the evaporation surface in the form of vapor. According to the experimental data, the thickness of the dehydrated layer was about two cm in specimens 20 cm deep at 60°C and 2.5 cm at 80°C. The migration of water slowed down as the distance from the surface increased because thermal gradients and moisture gradients between adjacent layers decreased. So the surface layer was dehydrated at about the same speed as specimens of the same depth and there was a danger that the surface zone would overdry.

It should be stressed, however, that evaporation decreases after the gradients level off or decrease and the outer layer is flooded with moisture carried up from the core. In this case, even if temperature is constant, thermodynamic forces try to level off the potential of the moisture transfer that was upset when the surface area of concrete was overdried. So the moisture condition of the surface layer in thicker specimens (and thus in actual structures) will be better as the concrete is cured further. But the increased dehydration of this layer at the important hardening stage when the concrete structure is formed is undesirable and results in deterioration of the quality and lower durability of the structural member.

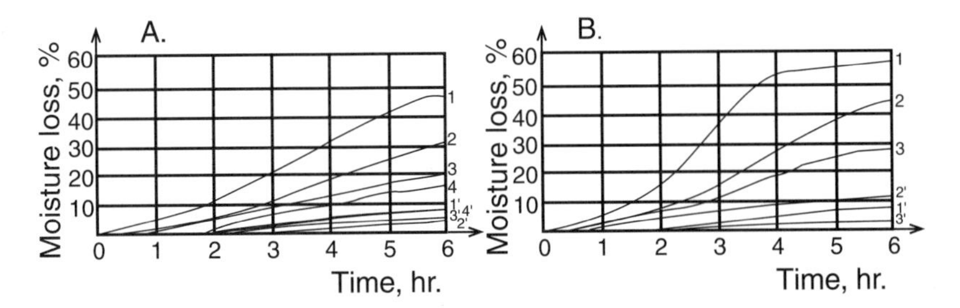

FIGURE 7.21 Evaporation of moisture from lightweight concrete specimens with different moduli of the exposed surface in the process of heating. a — heating at t = 80°C and b — heating at t = 98°C. 1 to 4 — exposed specimens; 1′ to 4′ — covered specimens, 1, 1′ — specimens with M = 40; 2, 2′ with M = 20; 3, 3′ — with M = 10; 4, 4′ with M = 5.

An analysis of the moisture evaporation kinetics shows that the process is most intensive at the end of heating and at the start of isothermal curing. Later on its rate does not change much.

Studies of the loss of the moisture in specimens made of lightweight concrete (with expanded clay) disclosed that it had the same relationship with the exposed-surface modulus but the amount of evaporated water was 1.5 to two times lower. The reason must be the considerable absorption of water by the porous aggregate and the appearance of discontinuites in water filling the cappilaries in the concrete. The discontinuites make transportation of water up to the surface much more difficult, which is manifested by the end results.

The covering of the exposed surface of concrete even with one layer of plastic film reduces the moisture loss many times close to an acceptable level. But this vapor barrier alone is not enough in winter for the exposed surface, particularly for thin structural members, and a heat insulator must be always laid on top of the vapor insulation to reduce the thermal gradient between the concrete and the environment.

The effects of the vapor insulation of the exposed surface on the loss of moisture. All the existing guidelines for the thermal treatment of concrete require that the exposed surface of a structure must be covered to prevent an excessive moisture loss. Studies at the NIIZhB found that the evaporation of moisture dropped tenfold in concrete heating, including the electrode method, when concrete specimens were covered with one layer of plastic film, as compared with the same specimens but having exposed surfaces (Figures 7.19 and 7.20).

When the surface is covered carefully, i.e., when the vapor insulator tightly envelopes concrete over the whole area, the moisture loss does not exceed 4 or 5% at moderate heating temperatures (50 to 60°C) and may reach 8 to 10% of the initial water content under higher heating temperatures and strong wind.

Observations established that thin flexible films, the most convenient for the vapor insulation (e.g., polyethylene or polyamide), let a small space of one or two mm saturated with water vapor form between concrete and the film even at the early heating stage. The pressure of the vapor-air mixture in this space is a little higher than normal, humidity is 100%, and temperature is close to that of the surface of concrete being heated, i.e., much higher than the ambient temperature. This interlayer is very useful because it (1) practically eliminates the moisture gradient between the concrete and the environment and creates an equilibrium of moisture, (2) decreases the thermal gradient between the concrete and the environment because of the higher temperature and acts as a buffer, and (3) also serves as a heat insulation layer to some extent between the concrete and the environment due to its lower coefficient of thermal conductivity as compared with concrete.

Unfortunately, these wonderful features of the space between concrete and film can be lost to a great extent if the concrete is not covered carefully or the film is rigid. We can improve the situation and upgrade the effectiveness of the vapor insulation cover, including that of a flexible film, by making a heat insulation layer over the vapor barrier from mineral wool, foamed plastic, or dry sawdust. The heat insulation decreases the thermal gradient and creates more favorable conditions for decreasing the moisture gradient in the gap between the vapor insulation and the concrete surface. Other conditions being equal, concrete specimens covered with a heat insulator on top of a vapor insulation lost 1.5 to two times less moisture as compared with specimens covered only with a free-lying vapor insulation layer in all experiments.

It is obvious from the above how important is the vapor insulation in concrete heating. When placed on the exposed surface, it protects concrete against excessive lost of water in erecting reinforced concrete structures at any ambient temperature and humidity. A small moisture loss that still takes place (4 to 10%) does not affect the quality of concrete and, on the contrary, is useful because it makes the concrete denser at the early hardening stage.

The effects of the heat-up rate and heating temperature on the loss of moisture. It is usually assumed that the loss of moisture increases when temperature of isothermal curing rises as the concrete is heated. But it is not only the qualitative aspect that is important but the quantitative aspect as well. The NIIZhB investigated the moisture loss in concrete for heat-up rates of 20, 60, 120, 240 and 500°C/hr, which produced time periods of three and one hours, 30, 15 and seven minutes, respectively. The experiments have shown that the concrete heat-up rate affects considerably the loss of moisture. The amount of moisture that evaporated as concrete was heated up decreased with an increase in the heat-up rate (Figure 7.22). It is quite natural because the volume of evaporated moisture depends on the time during which the temperature difference between the concrete and the environment was maintained. Consequently, accelerated electric heating of concrete is beneficial (this refers, of course, to electric preheating of concrete mixture).

Another regularity was found by experimental investigations, namely: as the heat-up rate was raised to a certain limit (it was 240°C/hr in the experiments), the amount of moisture that evaporated from concrete at the isothermal curing stage increased. It went down when the rate of rise in temperature was increased further. This phenomenon was observed in normal weight concrete and in lightweight concrete alike.

The reason for this variation in the moisture loss may be the following: when concrete mixture is heated up very quickly, the expansion and the outflow of the vapor-gas mixture are almost unimpeded. So there are no or very little (depending

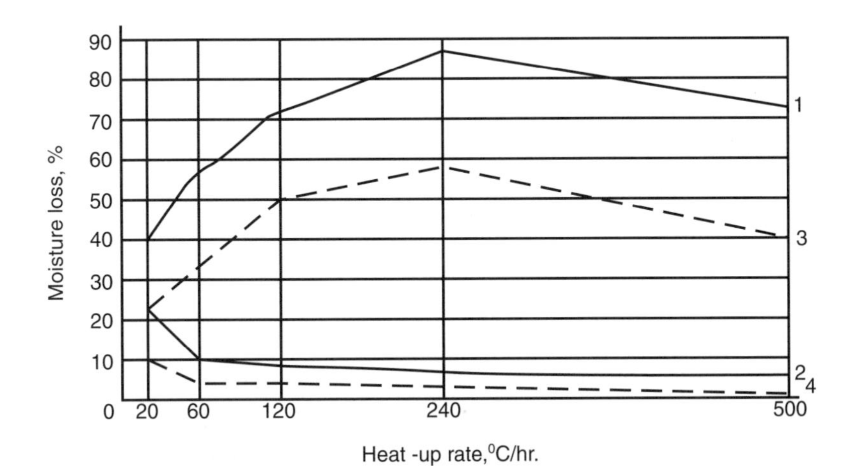

FIGURE 7.22 Evaporation of moisture from exposed specimens according to the rate of electric heating of concrete up to 80°C. 1, 3 — evaporation after seven hours of heating, 2, 4 — evaporation during time of heating. ______normal weight concrete, – – – – lightweight concrete.

on the rheology of a concrete mixture) internal cavities where moisture later evaporates and thus increases the amount of water vapor that can loosen the concrete and so adversely affect its structure and properties even more. The directivity of capillaries in the outflow of water vapor from concrete manifests itself less with rise in plasticity of the mixture because the channels are closed at once. This phenomenon can be roughly compared to boiling water when air bubbles formed at the bottom and coming to the surface do not leave any traces of the migration in the system.

A different picture takes place when the concrete mixture is heated up slowly or when a stiff mixture is heated up very rapidly. Little moisture is lost at the initial stage and the process accelerates only when temperature approaches the level of isothermal curing . The concrete mixture will have thickened by then, a certain, though small, structural strength will have formed and this becomes an obstacle to the free outflow of water vapor from the core to the surface. The vapor makes characteristic voids and cavities in the interior, noticeable visually, where water evaporates later on and the vapor damages the structure of the material as it tries to break through.

Concretes heated up at a medium rate have more directed capillaries — the migration paths of vapor to the surface — which cannot close because the concrete have attained a certain structural strength.

An analysis of the experimental findings showed that the absolute loss of moisture was high in specimens with exposed surfaces at any heat-up rate. It was 73 to 87% of mixing water for normal weight concrete and 39 to 53% for lightweight concrete at a rate ranging from 120 to 500°C/hr. Although the amount of evaporated moisture is low in isothermal curing when temperature rose very rapidly, it still is quite high and must not be allowed in actual concreting.

TABLE 7.4
Evaporation of moisture from concrete mixtures in the process of electric preheating and placing into a form

Type of concrete	Heat-up conditions	Amount of evaporated water in % of initial water content		
		During heat-up	During placing and compaction	Total
Normal weight concrete	Covered surface	0.5	7.3	7.8
	Exposed surface	2.1	5.9	8.0
Lightweight concrete	Covered surface	0.8	7.4	8.2
	Exposed surface	1.1	5.6	6.7

The moisture loss from concrete that was subjected to accelerated electric heating and placed when hot is of great practical importance. It was found by experiments that up to 8% of initial water evaporated when preheated concrete was placed (Figure 7.4). The amount of evaporated moisture depends to a great extent on how fast and in what fashion the concrete was placed into a form. Little water evaporates in the process of electric heating even if the mix was heated in an open bin.

A study of the relationship between the moisture loss from concrete and the temperature of isothermal curing proved that the evaporation increased with rise in temperature. We can assume with an accuracy sufficient for practical purposes that the increase in moisture evaporation is uniform and is about 1% per 1°C within the temperature range between 20 and 60°C and about 0.3% per 1°C at 60 to 80°C. So, at higher temperatures the increment of the moisture loss per each degree of rise in the temperature of isothermal curing decreases although the total amount of the loss grows.

As can be seen from the above findings, appreciable moisture losses occur in heating concrete with an exposed surface at all heat-up rates. So the thermal treatment of a structure should be carried out after its exposed surface was covered with a vapor insulator regardless of the temperature of isothermal curing and the rate at which it was reached.

The effects of wind velocity on the loss of moisture. Heat and mass exchange with the environment of a concreted structure in the process of heating accelerates under the action of wind that aids the removal of heat and moisture from the concrete surface along with convective flows that move air masses upwards. In laboratory tests, heated specimens were subjected to wind velocities of five and 10 m/sec.

The test demonstrated that the amount of evaporated water increased fast with wind velocity from any concrete at various heating temperatures (Table 7.5). The movement of air in parallel to the exposed surface creates a low pressure at the interface of the media. Partial pressure above the concrete surface drops too due to the rarefaction and since the boundary layer gets thinner because of a fast supply of fresh portions of air less saturated with water vapor to the surface. The result is the appearance of the moisture transfer potential and a faster inflow of water from the concrete core to the surface.

TABLE 7.5
Evaporation of moisture from concrete in electric heating for different wind velocities

Type of concrete	Heating temperature in °C	Wind velocity in m/sec	Water content of specimens in grams	Evaporation of water on exposed surface of specimens in grams
Normal weight concrete	60	0	366	119
		5	295	184
		10	287	212
	80	0	367	147
		5	264	186
		10	283	226
Lightweight concrete	80	0	362	108
		5	332	171
		10	382	235

It was found that if the evaporation surface of concrete was covered with one layer of plastic film, the moisture loss due to wind dropped three or fourfold. So, to sum up, we can say that in electric heating the direction of the moisture flow coincides with the heat outflow from the start of the thermal treatment until it ends increasing sharply with an increase in the temperature gradient.

The elementary precautions that consist in covering the exposed concrete surface with a vapor insulation layer preserves water in the material, which is needed for normal chemical reactions and structuring to produce high-quality and durable concrete.

7.7 ELECTRODE HEATING PROCEDURES

The organization of concreting with electrode heating depends on the type of plain or reinforced concrete structure to be erected, where this method may be used to accelerate the hardening of concrete. Due to specific features of the procedure whereby concrete is brought into electric circuit, not every structure can be heated successfully with the electrode method. It works well in heating nonreinforced or weakly and moderately reinforced structures of simple shapes. The electrode heating is unsuitable for heating concrete in structures of complex shapes, in thin-walled structures, and in heavily reinforced structures, including those reinforced with rolled steel sections, such as channels or I-bars. The reason is that it is difficult to heat concrete uniformly in such structures, it may be overheated in some spots, and there is a real danger of short-circuit of electrodes with steel and thus of the breakdown of the whole operation.

The above should be taken into account in concreting. Electrodes are chosen for a specific structure as best suited to its shape and reinforcement. A flow chart is made before the operation. It contains a schematic diagram of the structure to be heated, including the layout of the reinforcement and of steel inserts; the layout of electrodes and their connections; the description of fasteners for the electrodes and methods of their insulation from the reinforcement; the layout of temperature-measurement holes; data on the cross sections and lengths of power feeding wires and cables; data on vapor insulation and heat insulation of exposed surfaces; and other data pertaining to concrete and its strength characteristics.

Depending on the chosen method of electrode heating and type of electrodes, they may be installed before concreting or, on the contrary, after the concrete was placed and compacted. When the electrodes are installed, it is very important to keep the required distance between them and the reinforcing bars. Various spacers and fasteners made of hardened cement mortar or plastic are used for this purpose. They are attached to the reinforcement when electrodes are installed in a structure before concrete placing. String electrodes are tightened and fixed by suspending from rebars on hooks insulated with rubber or plastic tubes put on them.

Electrodes are inspected, their connections and wiring are checked. The main thing is to make sure that the electrodes are not displaced from their designed positions. The connections of the electrodes should be clean and be in tight contact with power leads.

Bronze caps are convenient for connecting power leads to rod electrodes. They are put on the ends of electrodes that protrude from concrete. It helps to connect the electrodes quickly for heating, to disconnect them after heating, and to use them repeatedly instead of connecting wires to the electrodes by twisting. Strip, band, and plate electrodes are best to be connected with screw clamps.

To avoid large heat and moisture losses in electric heating, try not to exceed the maximum permissible heating temperature (70 or 80°C), make sure that the dimensions of electrodes and their layout follows the design, and cover exposed surfaces with vapor and heat insulators.

When wooden panels with attached strip electrodes are installed on the surface of placed concrete, it is advisable that they be vibrated slightly for several seconds manually or with a vibrator to ensure the tight contact of the electrodes with concrete.

The exposed surface of a structure should be covered with vapor and heat insulation immediately after placing of concrete. If the surface is large, it is covered as the concreting proceeds. In severe frost the formwork has to be covered too to avoid a large heat loss. The formwork should be covered before concreting.

The temperature conditions and electric characteristics, such as the strength of current and voltage should be monitored all the time in the process of electric heating.

Careful preparations for the electric heating of concrete and its competent organization guarantee erection of reinforced concrete structures of high quality in a short time, and at minimum power costs.

8 Preheating of Concrete Mixtures

8.1 GENERAL CONSIDERATIONS AND THE PRINCIPLE OF THE METHOD

The method of electric preheating of concrete mixtures was proposed by Prof. A. S. Arbeniev and was a logical outcome of investigations directed towards expanding the range of the thermos method. As was pointed out above, thermos curing is the most cost-effective method of curing concrete in cast-in-place structures erected in winter. However, it is applicable only to the structures that have a low surface modulus under four to six because of the low temperature of concrete when it is placed in the formwork.

It was found in winter construction practice that the temperature of concrete discharged from a mixer with a temperature of 20 to 30°C usually does not exceed 10°C after transportation, handling, storage, placing, and compaction in a cold formwork with the reinforcement. The temperature of placed concrete properly insulated may rise due to the heat of hydration, but it may compensate with a vengeance the heat loss to the environment only in massive structures. It should be emphasized that a higher temperature of concrete produced by a batching plant does not help but makes concreting more difficult because the concrete loses plasticity very quickly.

Electric preheating of concrete mixtures on the spot near the placing site allows to avoid the high heat loss in transportation, handling, and other operations because the concrete can be taken from the batching plant at a temperature of 10 or 15°C in any frost. Concrete mixtures cool down to 0°C slower than, for example, from 30 to 10°C because of lower temperature gradients. The preheating of a concrete mixture at the site makes it possible to reduce considerably the consumption of fuel used for heating aggregates and water and to avoid special precautions needed to protect the concrete against cooling during transportation.

Fast placing of a heated concrete mixture minimizes the heat loss and helps to maintain its highest possible temperature at this stage of concreting. When placed in the formwork at a relatively high temperature (usually at least 50°C), the concrete is compacted and heats up by the heat of hydration to a higher temperature than a cold concrete due to a faster heat evolution when the concrete is hot.

Thus, with this method, concrete has a low temperature when delivered to the site but then it is heated up to the required temperature in a special bin. The concrete is then placed in the formwork quickly while it is hot, it is compacted, covered securely with a heat insulator to prevent it from cooling too fast, and is cured until it attains the strength as designed by the time of formwork removal and freezing. The simplicity, effectiveness, and economic advantages of the method soon made it quite popular. It was adopted rather widely in construction and in the precasting industry.

8.2 THE THEORETICAL BACKGROUND OF THE ACCELERATED HEATING OF CONCRETE MIXTURES

Accelerated electric heating of a concrete mixture to be then placed and compacted while hot can affect the microstructure and the macrostructure of the material just like any other method of thermal treatment.

Studies conducted at the NIIZhB demonstrated than the new hydrate formations in the concrete that was subjected to accelerated electric heating do not differ from those produced by any other concrete heating method. So only the rates of chemical reactions and certainly physical features of structuring may change. The hydration of binders usually lasts rather long under normal temperature conditions. So the acceleration of this process has a potential for shortening the time needed for the cement matrix to harden and become hardened cement paste.

The first stage in the reaction of cement with water, which is called the "kinetic stage", ends in the initial precipitation of new hydrate formations from the liquid phase they supersaturate. They settle on the surfaces of the binder particles as films. The hydration rate depends mainly on the rate of the direct chemical reaction between the cement clinker minerals and water at their interface. The hydration process is not complicated by diffusion phenomena and has a high activation energy. High temperature accelerates this process because the higher total kinetic energy of molecules increases sharply the number of active molecules and thus the number of effective collisions which intensify the chemical reaction.

P. P. Budnikov, S. M. Royak, Yu. S. Malinin, and other researchers found that the induction period in the hydration of cement becomes shorter with rise in the temperature of moist steam curing. The use of fast electric heating of concrete at the early hardening stage also relies on the chemical kinetics of the heterogeneous processes that take place during that period, including the inflow of the solvent to the surfaces of cement grains, the chemical reaction between water and cement at the interface of the phases, and the outflow of the chemical reaction products (new hydrate formations) from the surfaces of the cement grains to deeper areas of the liquid phase. The rate of the process depends on the rate of one of its stages, i.e., the diffusion phenomena that take place at the first and third stage. So the reaction follows the diffusion kinetics.

Dissolved substances, i.e., a more concentrated liquid phase, are removed from the surfaces of the cement grains according to the law of molecular diffusion due to the difference in the concentrations of the substances. This process is rather slow and can be accelerated only by raising temperature to increase the kinetic energy of the molecules of water and of the substances dissolved in it making them move faster.

Consequently, the sooner the temperature reaches high values the faster the cement clinker minerals dissolve due to the higher rate of diffusion. The electric current helps this process to some extent when it is passed through the concrete.

Rise in temperature in the heterogeneous reactions that follow the diffusion kinetics still accelerates hydration less than in the kinetic region. The reason is the formation of films around cement grains due to low diffusion coefficients of new hydrate formations in a supersaturated solution. The speed at which the shielding films form, the density of the films, and their adhesion to the unreacted nuclei of the cement grain increase with temperature. This results in their lower permeability that depends on the dispersion of the reaction products (the density increases and the permeability decreases with an increase in dispersion).

Since the thermal treatment promotes the appearance of more dispersed crystals of the new formations, we can expect the film forming phenomena to speed up under accelerated electric preheating of concrete. It may lead to an earlier termination of the intensive cement hydration despite an increase in the amount of the initial binder that enters into the chemical reaction at once at the early stage.

We may avoid or weaken this phenomenon by curing concrete for some time before heating at a low above-zero temperature. So it is desirable that this technique be a stage in the total cycle of the thermal treatment of concrete by this method.

Studies disclosed that, indeed, the degree of cement hydration decreases a little with the accelerated electric heating as compared with the thermal treatment whereby temperature is raised slowly, which is usually the case. But the decrease is only by several percents and does not make any difference.

The accelerated temperature rise helps to destroy the metastable films of calcium hydrosulfoaluminate that retard the hydration of cement. It should be emphasized that during the usual thermal treatment the disintegration of calcium hydrosulfoaluminate falls on the period when the structure strengthens and the processes whereby it is disturbed are irreversible and adversely affect strength characteristics of hardened cement paste. When the temperature is raised quickly, the films form and break at an earlier period of the structure formation and so the latter does not deteriorate, and even if it does, to a much smaller extent.

The accelerated rise in the temperature of a concrete mixture requires that the concrete be vibrated when hot. This technique is a must for accelerated methods of the electric thermal treatment of concrete.

It is well-known that repeated vibration of concrete mixtures at normal temperature favorably affects the strength of the material.

The vibration has an even greater impact when concrete is preheated electrically (Figure 8.1). Its benefits are twofold. It eliminates all kinds of structural defects of physical nature and favors faster hardening and formation of the crystalline skeleton of the future hardened cement paste. The impact on the chemical aspects of hardening manifests itself in the form of breaking the metastable films of calcium hydrosulfoaluminate by mechanical action and of providing favorable conditions for the formation of a stronger crystalline skeleton.

The films of the new formations are broken by vibration in a hot state as they are torn away from cement grains when the latter collide. This exposes unhydrated parts of the nuclei of the cement grains making them accessible for water. The remains of the films in the form of finely dispersed coagulation structures with some broken crystals partially dissolve but mostly reform into a system with a denser packing of particles. So the crystallization structure is made up mostly of crystals joined along cleavage planes and much less by their ingrowth at random contact points. This reduces stresses inside the crystalline skeleton and improves the strength of concrete.

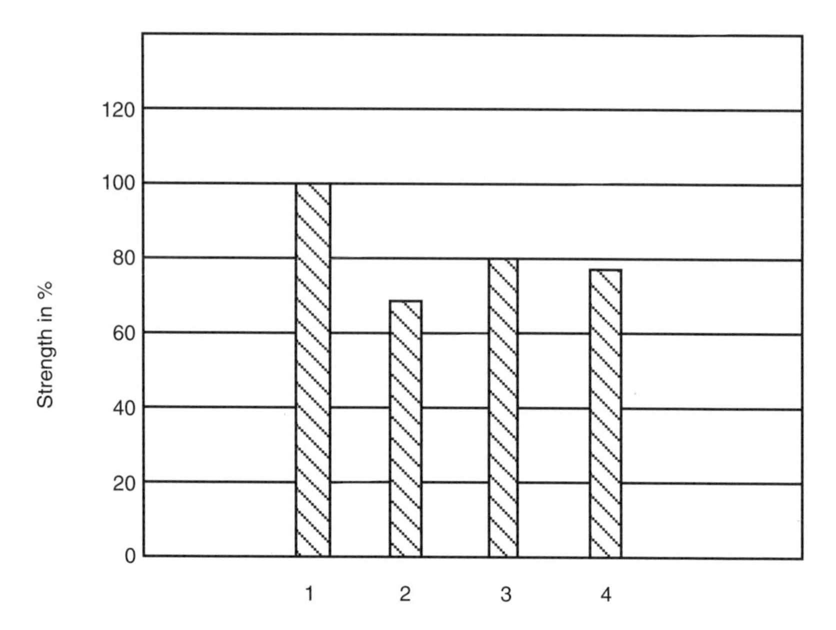

FIGURE 8.1 The strength of concrete after seven days depending on repeated vibration. 1 — electric heating up to 80°C for 15 minutes with repeated vibration, 2 — the same without repeated vibration, 3 — without heating, vibration repeated in 15 minutes after casting, 4 — the same without repeated vibration.

Electric current passing through a concrete mixture generates electric and. magnetic fields. The effects of the electric field and of the magnetic field on the finely dispersed component and on the liquid phase of concrete have been studied very little and specific information about the extent of these effects is not available. But there are theoretical grounds for this and the influence of the fields is not improbable.

Collisions of dispersed particles due to their electrophysical oscillations in electric heating break mechanically the films of new substances, which were formed, and provide a better access for water to the unreacted part of cement grains.

The polarization of water molecules intensifies their reaction with the cement clinker minerals, the dissolution of the latter, and the saturation of the liquid phase with hydration products. The polarization of colloidal and dispersed particles results in the formation of associates of regular (perhaps directed due to polarization) construction with stronger bonds of electric nature. The skeleton composed of such associates should reduce inner stresses of crystallization character and strengthen the microstructure of hardened cement paste.

The electromagnetic field generated in concrete by electric current reduces the viscosity of water due to the electrodynamic effect and thus enables the water to penetrate deeper into submicropores and cracks of cement grains and of the films of new formations accelerating the hydration of the binder even in the presence of calcium hydrosulfoaluminate films. Water undergoes certain structural changes in the electromagnetic field, which produce more crystallization grains and a greater coagulation in the supersaturated solution. The structure of hardened cement paste is composed of smaller crystals increasing its isotropism and strength. Finally, the effect of plastification that occurs under the action of the electromagnetic field on water should improve the rheological parameters of the material during vibration.

The foregoing, of course, is the most pronounced and perceptible when the electromagnetic field applied to concrete has optimum parameters. This problem has not been studied at all as far as the accelerated electric heating of concrete is concerned and it is difficult so far to evaluate the importance of the field.

Accelerated electric heating accompanied by vibration of hot concrete improves the microstructure of hardened cement paste and reduces the inner stresses that arise in the process of crystallization structuring. It was demonstrated by many researchers and there is no need to dwell on it.

The accelerated electric heating of concrete has certain advantages over thermal treatment methods with a gradual rise in temperature from the standpoint of the thermodynamics of irreversible processes. According to Le Chatelier's principle, cooling makes exothermic reactions more complete. So it is better to raise temperature in concrete before intensive heat evolution, which helps to have a steep rise of the exothermic curve as soon as possible and stops the inflow of external heat while it evolves rapidly inside. Then, during the period of thermos curing, heat can be removed naturally from placed concrete. When temperature is raised gradually during a thermal treatment, concrete receives maximum heat during the period of the greatest heat evolution from hydration, which should damp out exothermic reactions. This may be one of the reasons why we cannot get a strength higher than 70 or 75% R_{28} by heating within a specified cycle.

This concise theoretical analysis of physico-chemical phenomena that occur during the accelerated electric heating of concrete shows that this method has a real scientific basis and can produce the material with the same structure as that of concretes that hardened under normal conditions. So we can recommended it for wide use in construction and continue research in this field.

8.3 INTERNAL HEAT EXCHANGE DURING THE ACCELERATED ELECTRIC HEATING OF CONCRETE

Internal heat exchange is particularly characteristic of the accelerated heating of concrete and is possible only under electrode heating. This mode of operation whereby concrete is heated up to the required temperature within five to 15 minutes is widely adopted for electric preheating of concrete mixtures, for accelerated heating in a form followed by another vibration, and for electric heating in the process of vibratory casting of prefabricated products.

The practice of accelerated electric heating have seen cases where the concrete cooled down considerably after current was cut off. The reason was redistribution of heat inside the nonuniform multicomponent system, that is the concrete mixture. Electric power is directly converted to thermal energy during electric heating only in the conducting component, i.e., cement paste. The cement paste heats fine aggregate particles almost instantly (the delay does not exceed several seconds). Coarse aggregate heats up slower depending on its type and particle size. The average temperature of the concrete drops as a result.

Considering the great theoretical and practical importance of the internal heat exchange during the accelerated electric heating of concrete, the NIIZhB initiated detailed studies of this problem. The results of the experiments show (Figure 8.2) that the diameter of a particle is very important for the time it takes to heat it. As can be seen in the diagram, the difference grows between the rise in the temperature of cement paste (upper curve) and the temperatures in the center of granules of expanded clay as their diameters increase. The kinetics of the aggregate particle heating is characteristic. For any size under 10 mm the temperature in the center of the particles lags behind that of cement paste by about the same value during almost the entire period while the temperature rises. This lag is 5°C for five mm particles and 10°C for 10 mm particles at the same specified heating rate. By the end of electric heating there is no difference in the temperatures of the paste and of the five mm particles because the cement paste reaches the require electric heating temperature 15 sec ahead of schedule . In the particles of 10 mm in diameter the maximum temperature is achieved in about one minute after the end of heating.

The kinetics of heating of larger particles is noticeably different from that of cement paste heating and the temperature difference between them grows as the size of the granules increases. Typically, an appreciable rise in the temperature of particles larger than 30 mm starts after it rose in the paste to 40 or 50°C. Further on, as the internal thermal gradient rises in the material, the aggregate heats up by conduction faster and at practically the same rate as the paste. But the temperature difference during the main heating period between the center of 30 mm particles and the cement paste is 30 to 35°C reaching 45°C for a particle size of 40 mm. In dry particles of expanded clay of this diameter the maximum temperature in their centers is reached within five to 12 minutes.

It is important that the total temperature of the mixture drops after electric current is cut off due to the internal redistribution of heat which is absorbed by the aggregate.

A similar picture is observed in heating concrete with damp aggregate particles. But because of their higher coefficient of heat conductivity the temperature lag at the main heating stage does not exceed 30°C even for 40 mm particles. The temperature already levels off in the system in three minutes after the current is cut off.

Under accelerated heating conditions for normal weight concrete with a dense aggregate of the same size the particle heating temperature lags behind cement paste by about the same value as in heating a water-saturated expanded clay. It depends on the density of aggregate and ranges from one to three minutes after the current is cut off. When a concrete mixture containing aggregates of different densities was heated for four minutes, the difference between heating rates of particles of different diameters could be seen quite well (Figure 8.3). Allowing for the coefficients of heat conductivity (12.5×10^{-4} m^2/hr for expanded clay, 22.7×10^{-4} m^2/hr for limestone, and 48.7×10^{-4} m^2/hr for granite), we can say that the higher was this coefficient for an aggregate material the faster its particles heated up. Obviously, the average value by which an aggregate is underheated, i.e., by which its temperature is lower than that of the liquid part by volume, is of great interest for selection of electric heating conditions because it is proportional to the quantity of heat to be absorbed

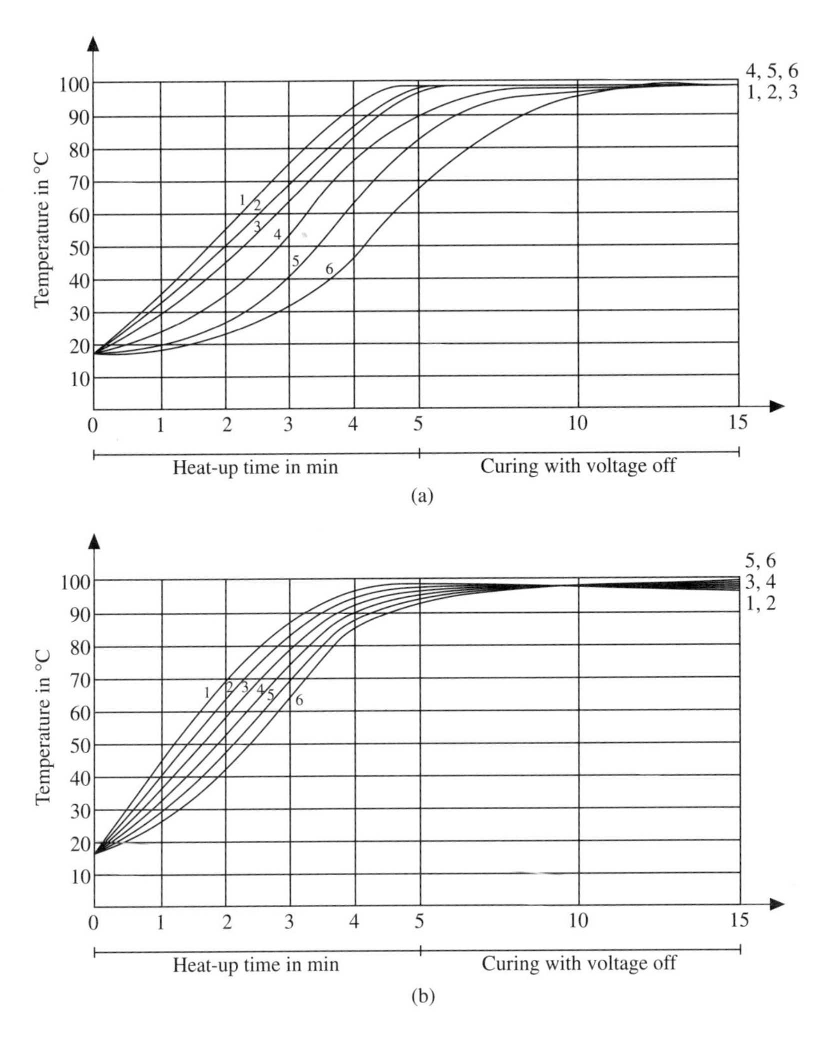

FIGURE 8.2 Variation in temperature in the core of expanded clay particles in accelerated heating of concrete mixtures. a — grains placed in mixture six minutes before heating, b — the same 45 minites before heating, 1 — temperature of the liquid part of concrete, 2 to 6 — temperature of expanded clay grains of five, 10, 20, 30, and 40 mm in diameter.

by the aggregate from the liquid after the end of heating to bring down the average temperature of the concrete mixture.

The dependence of the underheating value on the aggregate size and on the rate and time of heating of the concrete mixture can be conveniently found in practice in a nomogram (Figure 8.4). The nomogram contains a family of curves, each corresponding to a certain concrete heating time, for each aggregate size. Aggregate is underheated considerably when the heating rate is high and the aggregate particles are large. When the aggregate particle size is under 30 mm, the drop of the temperature of the concrete mixture after the end of heating is mostly within 5°C. The

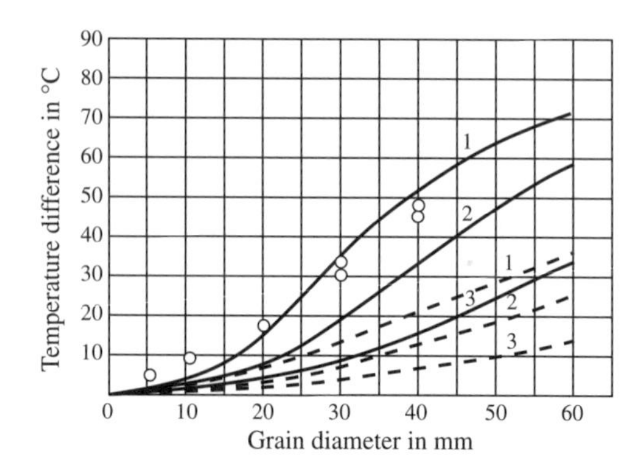

FIGURE 8.3 The temperature differences in the liquid part in particles of coarse aggregate at the end of four minutes heating of concrete mixture at a rate of 1125°C/hr. 1 — expanded clay, 2 — limestone, 3 — granite. ________ temperature in the core of grains, _ _ _ _ _ mean temperature over the volume of grains.

kinetics of the heating of coarse aggregate in concrete, the effects of certain factors on it, and the drop of the temperature of the concrete after the end of electric heating can be described analytically under certain assumptions. But this calculation is quite complex and is not made here.

8.4 CHANGES IN PLASTIC PROPERTIES OF CONCRETE CAUSED BY ACCELERATED HEATING

The plastic properties of concrete mixtures are one of the most important characteristics for the selection of vibratory mechanisms and of conditions of compaction. They change greatly when the concrete mixture is heated and this should be taken into account in concreting. Studies proved that the loss of plastic properties in electric heating was affected by a number of factors, including the mineralogical composition of cement used, the quantity of water that the concrete mixture contains, the presence of plasticizers, the rate of rise in temperature and its final value, etc.

It was found that an increase in the content of C_3A in cement greatly reduced the setting time with heating. Hot concrete mixtures with belite cement, on the contrary, retain their plastic properties much longer. Concrete mixtures with a high W/C and an increased initial slump have a quite satisfactory workability in a heated state. Plasticizers help to ensure acceptable properties of hot mixtures for concreting. The rate of the temperature rise has an important bearing on rheological characteristics of concrete mixtures. These characteristics change to a much smaller extent when the temperature rises rapidly than with slow heating. Conversely, the increase in the stiffness of a concrete mixture is directly related to the heating temperature.

The plasticity of a concrete mixture can be changed by the loss of moisture due to evaporation in the process of electric heating, during transportation, and in handling the hot mix, which may make placing and compaction of concrete considerably more difficult if the concreting is not properly organized.

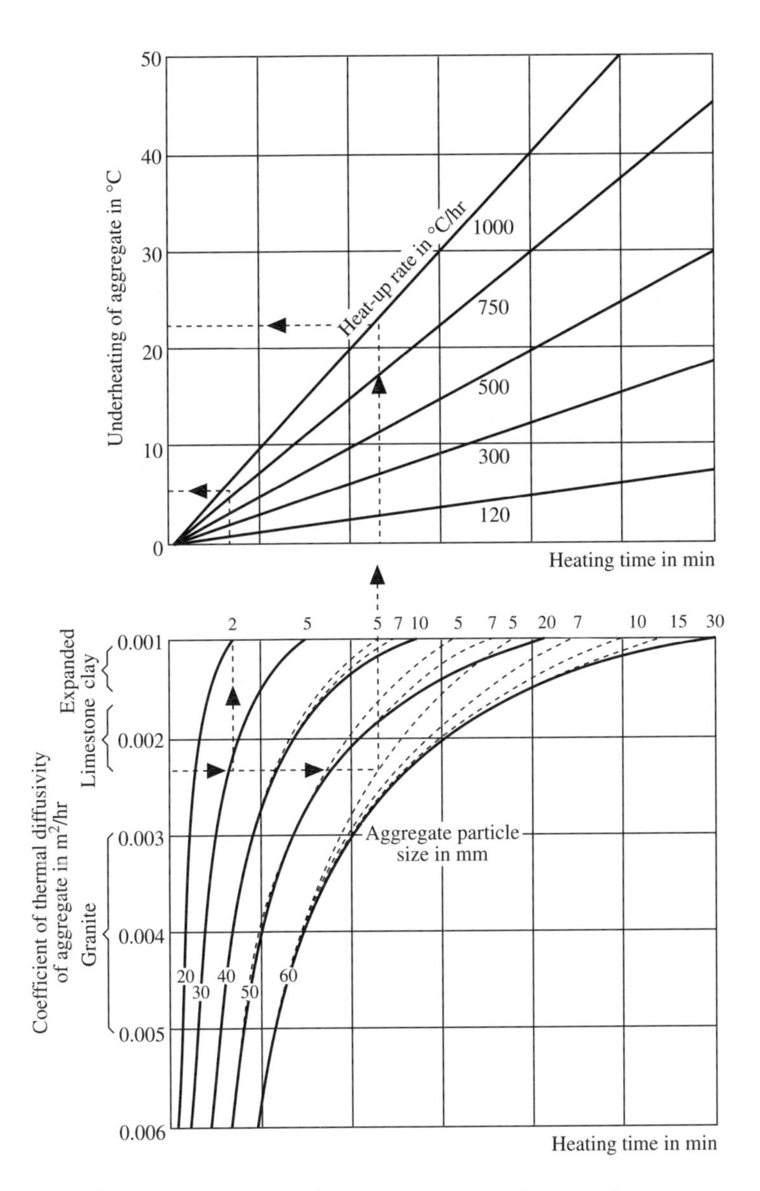

FIGURE 8.4 The nomogram used to determine the average heating of coarse aggregate for accelerated electric heating of concrete mixtures.

A study of the variation of plastic properties of concrete mixtures under accelerated electric heating disclosed basic causes of this phenomenon, namely:

- A change in the hydrophilic properties of aggregate particles whose capillary water absorption increases with rise in temperature due to the drop of water viscosity in the range between 10°C and 80°C.
- A change in the hydrophilic properties of cement grains.

- An increase in the chemical activity and reactivity of water as the temperature rises and the resultant increase in the kinetics of its binding in new formations.
- A decrease in the amount of free (mechanically bound) water in the concrete due to its accelerated chemical and physical binding and loss by evaporation to the environment.

Experiments have shown that the water absorption of coarse aggregate increased with temperature according to its structure and type of porosity (limestone, crushed brick, expanded clay, etc.). It is 2 to 7% higher at t = 80°C than at t = 20°C. The viscosity of water drops by about two-thirds in the same temperature range due to disassociation of aggregations and their high mobility. The appearance of a large number of free molecules and of much smaller associations than at t = 20°C enables water to penetrate into smaller pores, capillaries, and microcracks in aggregate particles and thus less water is left in the intergrain space. The same can be said about the penetration of water into cement grains, which affects the need of the cement for water in the hot state.

The increased chemical activity and mobility of water molecules increases the number of contacts of the water with cement clinker minerals and accelerates the diffusion processes that bring new portions of water to the grain surface and take away dissolved hydration products to the space between the grains. All this accelerates the reaction of water with cement, which is well demonstrated by the setting time variation.

To determine the setting time of cement at different temperatures, the NIIZhB designed a special device (Figure 8.5) based on the same principle as the Vicat needle. A form with cement paste specimens is placed into a special chamber made of acrylic plastic, where the required temperature and 100% humidity are maintained. The latter prevents evaporation of moisture from the specimen since it affects to a great extent the setting time which is thus determined only according to the temperature factor. Tests made in this device proved that the time between the start and the end of cement setting is shortened dramatically as temperature rises.

When water is bound faster chemically in new hydrate formations and physically with molecular forces on submicrocrystals of hydration products whose integral surface grows quickly due to intensive crystallization, the amount of free water in the material also decreases. Evaporation to the environment caused by the high temperature and moisture gradient also reduces free water. The loss of moisture in the process of electric heating of the concrete mixture and as the concrete is placed may be as high as 5 or 7%. It was observed that the plasticity a hot concrete mixture with dense aggregates had at the end of electric heating decreases much longer when there was no evaporation of moisture.

Tests of the workability of concrete mixtures at different temperatures found that it depended on the latter and the curing time in the hot state (Table 8.1). A concrete mixture (t = 98°C) with Portland cement from the Voskresensk Cement Factory with a medium content of C_3A became so stiff that it was practically impossible to place. Concrete with another Portland cement from the Bryansk Factory, which contained little C_3A, remained plastic even at the same temperature,

FIGURE 8.5 A device for determining cement setting time for different heating temperatures.

but after 15 minutes of curing at this temperature its stiffness grew sixfold and after 30 minutes tenfold. It should be emphasized that curing of a hot concrete mixture for 15 minutes at various temperatures increases its stiffness at least 1.5 or two times and for 30 minutes two to 10 times on the average (depending on the curing temperature and the mineralogical composition of cement). Adding 0.2% plasticizer — sulfate liquor stillage — by cement weight can reduce stiffness 1.5 to two times in all cases.

The type of aggregate affects considerably the variation in the stiffness of a concrete mixture. While concretes with dense aggregates (granite or dense limestone) lose plastic properties increasingly during the whole heating period, the plasticity of concretes with porous aggregates (expanded clay or volcanic slag) increases in the process of heating but then drops sharply after the required temperature is reached and curing continues at this temperature. The reason for this interesting phenomenon is as follows.

As the temperature rises, the air entrapped in the aggregate pores expands and starts to drive the absorbed water out into the intergrain space. Vapor produced by the internal evaporation of water in the pores of the aggregate particles also helps because it increases the amount of the gaseous phase inside the aggregate granules. The space between the particles is increasingly filled with water as the temperature rises and the mixture becomes more workable. This may cause concrete mixtures with porous aggregates to separate if their composition was not properly designed. The water goes down and flows out if the gate of the bin is not tightly closed.

When the concrete mixture reaches a specified temperature level and the heating stops, the reverse process — the saturation of aggregate particles with water —

TABLE 8.1
The variation of the plasticity of concrete mixtures according to the electric heating temperature and curing time at this temperature

Heating temperature in °C	Curing time of hot mixture at heating temperature in min	Stiffness of hot concrete mixture for initial slump of 3 to 5 cm in sec		
		Portland cement No. 1	Portland cement No. 2	Portland cement No. 3
40	0	$\frac{35}{25}$	$\frac{10}{7}$	$\frac{20}{10}$
	15	$\frac{43}{32}$	$\frac{19}{10}$	$\frac{30}{17}$
	30	$\frac{65}{40}$	$\frac{21}{13}$	$\frac{40}{24}$
60	0	$\frac{95}{60}$	$\frac{12}{9}$	$\frac{25}{23}$
	15	$\frac{180}{125}$	$\frac{19}{12}$	$\frac{60}{43}$
	30	$\frac{230}{215}$	$\frac{29}{15}$	$\frac{150}{72}$
80	0	$\frac{155}{80}$	$\frac{14}{11}$	$\frac{35}{33}$
	15	$\frac{215}{170}$	$\frac{43}{14}$	$\frac{120}{63}$
	30	$\frac{340}{190}$	$\frac{70}{27}$	$\frac{210}{130}$
98	0	$\frac{140}{110}$	$\frac{16}{13}$	$\frac{70}{27}$
	15	$\frac{280}{135}$	$\frac{90}{28}$	—
	30	$\frac{295}{115}$	$\frac{164}{70}$	—

Note: The numerator shows stiffness without admixtures, the denominator stiffness with a plasticizer — sulfate liquor stillage — in the amount of 0.2% by cement weight; Portland cement No.1 contained 8% C_3A, Portland cement No. 2 contained 6% C_3A, and Portland cement No. 3 contained 7% C_3A.

begins. It goes faster when the temperature is maintained at the same level than at normal temperature because the viscosity of water is lower. The intergrain space dehydrates faster as a result and the mixture soon becomes quite stiff.

However, at the site, the temperature always goes down a little because of the internal and external heat exchange after voltage is cut off. Having decreased in volume much more than the other components, the vapor and gas mixture that remained in the aggregate pores creates a lower pressure and water is sucked rapidly into the particles. When concrete separates and water moves down into the lower layers, the upper layers lose their workability even faster.

When a heated concrete mixture with porous aggregates is cured at a high temperature, its stiffness grows faster than that of concrete with dense aggregates. Thus, concrete mixtures with any cement lose plasticity and turn into a loose mass after 30 minutes at t = 98°C.

The brief analysis made above and experimental data on measurements of the plasticity of concrete mixtures under accelerated electric heating shows the great importance of this factor and the need to allow for it in electric heating at the site. It is usually done by adding 8 to 15% more water than designed or by introducing a plasticizer.

8.5 PHYSICAL-MECHANICAL PROPERTIES OF CONCRETE SUBJECTED TO ACCELERATED HEATING

The methods used to accelerate hardening are mostly aimed at concrete achieving a high compressive strength at the shortest time possible. But reinforced concrete structures are not always subjected only to compression. It is very important for their different types to have also crack resistance, frost resistance, impermeability to water, or resistance to various corrosive media in service. So, speaking about accelerated hardening, we must bear in mind the need to create conditions for the formation of an optimum structure of concrete in the shortest time possible to ensure its most important physico-mechanical characteristics that allow structural members to work under various service conditions. Remember that basic properties of concrete depend on its composition and materials used. The hardening conditions should be favorable for the structuring to let the concrete withstand the loads and environmental actions for which the member is designed. The accelerated heating of concrete used to make it harden faster should meet the same requirements.

The NIIZhB has conducted extensive research to study the whole set of properties of the concrete that was subjected to accelerated heating. The investigations included tests of normal weight concrete and lightweight concrete with expanded clay, which contained various types of cement. The kinetics of strength development was compared with that of concretes of the same compositions but steam cured at 60, 80, and 95°C. The specimens were tested two hours after forms were removed.

As could be expected, the kinetics of concrete strength development both under accelerated heating followed by vibration and under steam curing depended on the mineral composition of cement at different heating temperatures. Thus, normal weight concrete with low-aluminate Portland cement from the Belgorod Factory (C_3A = 4%) developed its strength more rapidly than medium-aluminate Portland cement made by the Gigant factory (C_3A = 8%) at 95°C. When the heating temperature was 80°C, the rate of the strength development by concretes containing the two cements was about the same, but when they were heated to 60°C, the strength of the medium-aluminate Portland cement concrete was a little higher than that of the low-aluminate Portland cement after the same period of isothermal heating.

Thus, the nature of strength development depends on the mineral composition of cement according to the heating temperature. Low-aluminate cements are most suitable

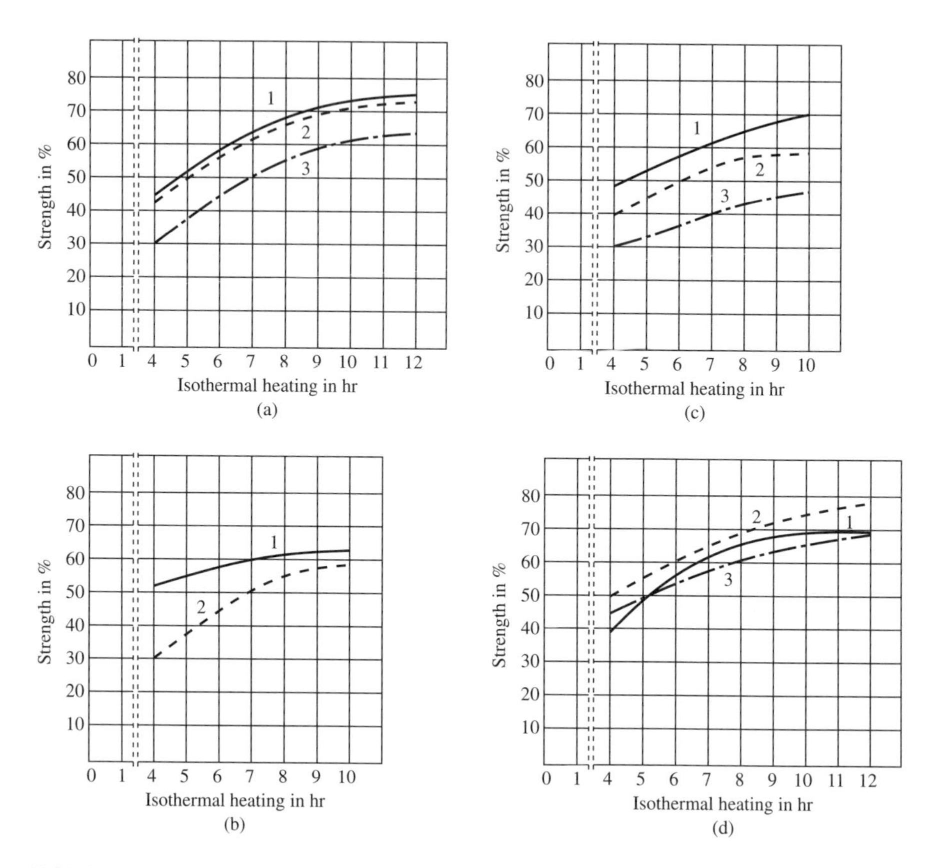

FIGURE 8.6 Strength development of lightweight concrete under accelerated electric heating according to the heating temperature. a, b, c, and d are concretes with low-aluminate Portland cement, Portland blast-furnace cement, rapidly hardening Portland cement, and average-aluminate Portland cement, respectively. 1 — at t = 95°C, 2 — at t = 80°C, 3 — at t = 60°C.

for concrete to be heated at a temperature higher than 80°C and medium-aluminate cements for temperatures lower than 80°C (Figure 8.6). Portland blast-furnace cement concretes are better to be heated at a high temperature level of 80 to 95°C. All this confirms the need to restrict the use of concretes containing cements with a high percentage of C_3A for accelerated heating at a temperature higher than 80°C just like for any other type of thermal treatment.

The studies have shown that ordinary Portland cements and high temperatures are the most effective means in accelerated heating if we want to obtain the highest possible strength immediately after the thermal treatment of the concrete. Concretes containing such cements can attain 70% of the their grade values after they were heated to 80°C. Portland blast-furnace cement concretes harden slower and rarely develop the 70% grade strength even after 10 hours of isothermal curing. Similar results were produced by steam curing.

An analysis of the above findings has revealed that concrete developed strength faster when it was subjected to accelerated heating than to steam curing. It happened to practically all concretes with different cements regardless of the temperature of the thermal treatment.

There was not much difference in strengths of concretes that underwent accelerated heating or steam curing after 28 days. But still the strength of concretes that were heated fast electrically and then compacted when hot was somewhat higher than of steam-cured ones. The concretes with accelerated heating either developed the same strength as similar concretes that hardened under normal conditions or the strength was lower but very little.

An interesting case is the strength development of concrete that was subjected to accelerated heating but then hardened under normal conditions. Of course, the heating temperature and the mode of curing affect it. The strength development rate after the heating and vibration of hot concrete was found to be dependent on cements used and on the heating temperature. Normal weight concretes with the low aluminate Portland cement from the Belgorod Factory continued to harden rapidly after the thermal treatment. Their strength exceeded the grade value by 10 to 30% after 28 days and by 10 to 40% after 90 days. Concretes with the medium aluminate Portland cement made by the Gigant factory developed their strength somewhat differently. The strength increased little — by 20 to 25% — as compared with its values immediately after heating at 80°C on the 28th and 90th day, i.e., it did not even reach grade strength in some cases. But when heated at 60°C, concretes with this cement continued to harden rapidly enough and exceeded grade strength by 10 to 20%.

The studies have shown that the higher the temperature of concrete heating and isothermal curing, the slower is the strength development afterwards. A similar picture was observed for Portland blast-furnace cement concretes and concretes with rapid hardening cement although the level of high temperatures was raised a little. Thus, when the heating temperature was 80°C, the strength of concretes with these cements not only attained grade strength at the age of 28 days but even exceeded it by 10%. But if concretes were heated at 95°C, their strength was underdeveloped by 4 to 12% (Figure 8.7).

Similar results were obtained in studies of lightweight concretes with expanded clay aggregate and the same cements.

The above findings show the negative response of rapid hardening Portland cement to heating at a high temperature. And we can draw a general conclusion from the above research that, although thermal treatment at a temperature higher than 80°C results in quite high initial strengths of concretes, the hardening may be quite slow later on and the strength may be underdeveloped on the 28th day. So it is not recommended to subject concrete to accelerated heating or to another thermal treatment at temperatures over 80°C. An optimum heating temperature for concretes with most Portland cements should be within 60 to 80°C range depending on the mineral composition of cement. For concretes containing highly active Portland cements the heating temperature at the level of 60°C is quite sufficient.

In addition to compressive strength, other properties of concretes that underwent accelerated heating are also important. Prism strength, modulus of elasticity, bending tension, and frost resistance were studied.

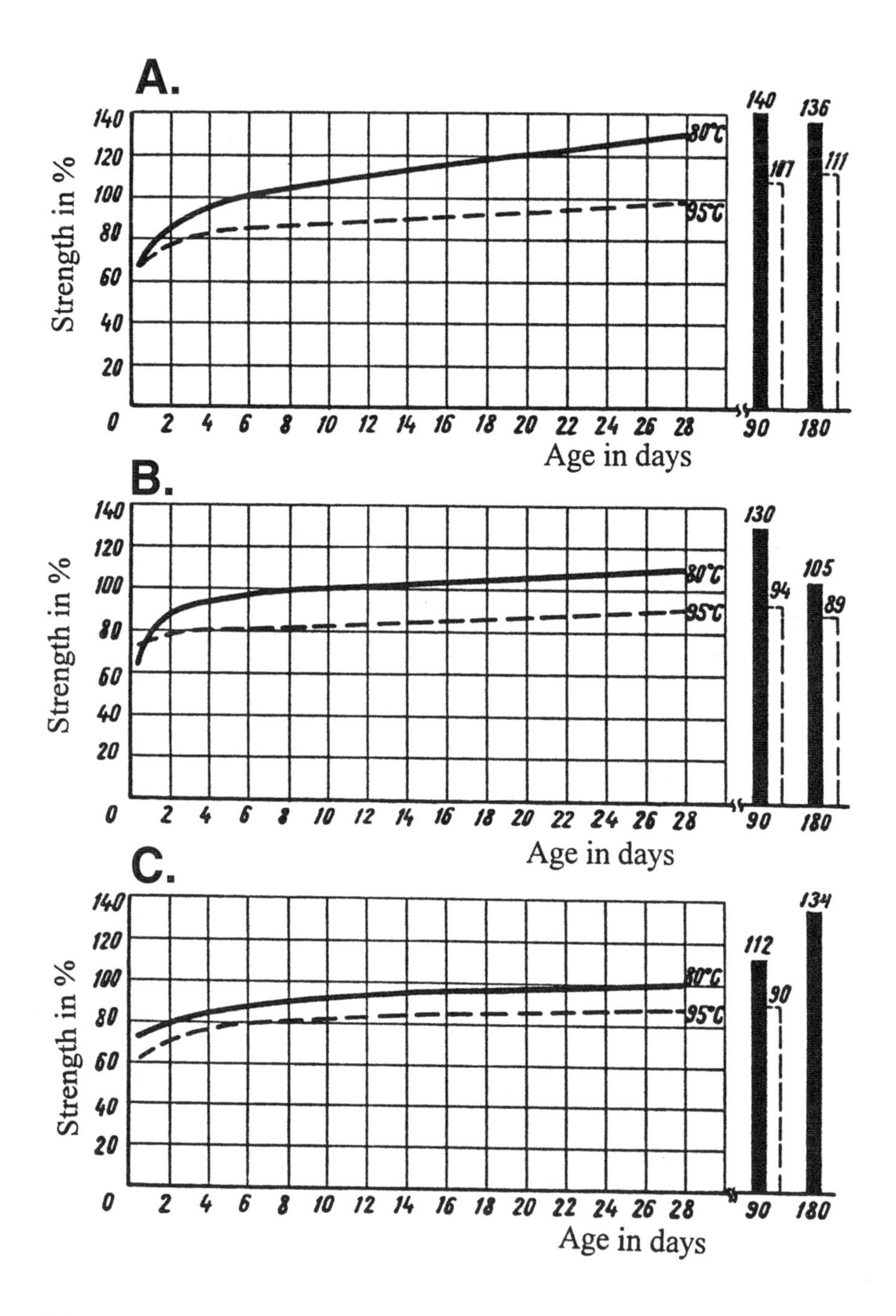

FIGURE 8.7 Strength development of normal weight concrete with various cements after accelerated electric heating. A, B, and C are concretes with low-aluminate Portland cement, Portland blast-furnace cement, and rapidly hardening Portland cement, respectively.

The prism strength of these concretes was practically the same as that of similar concretes that hardened under normal conditions. The strength was underdeveloped in some cases but in other cases it exceeded the value attained under normal conditions. So we can say that on the average accelerated heating does not adversely affect this type of strength.

The situation was quite different with modulus of elasticity. It was 11 to 15% lower in concretes with all types of cement as compared with concretes that hardened under normal conditions. It should be noted that this decrease in modulus of elasticity is generally characteristic of the concretes that were subjected to a thermal treatment. Some researchers even recommend to reduce the modulus of elasticity by 25% in designing reinforced concrete structures to be subjected to thermal treatment. This recommendation is unjustified and is excessively strict but nevertheless the modulus of elasticity does decrease in heated concretes. A comparison of the modulus of elasticity as well as of prism strength with normative values of these characteristics for concretes of similar grades (data of the Building Code) shows that they exceed the normative values by 2.5% to 18% according to the type of cement used.

The tensile strength of concretes that underwent accelerated electric heating and compacted immediately after placing by vibration was also studied. The tensile bending strength of the concretes was practically no different from that of concretes that hardened under normal conditions regardless of the type of cement (Portland blast-furnace cement concretes were tested) at the age of 28 days. It even exceeded the strength of the latter on the 90th day probably because of increased density.

Similar results were obtained by studying the bond between normal weight concrete and steel. The bond between lightweight concretes with expanded clay and steel was even higher.

Frost resistance is an important property of concrete, which determines its durability. This characteristic is known to depend largely on concrete curing and the thermal treatment conditions. An ordinary thermal treatment of concrete, e.g., steam curing, somewhat reduces frost resistance because of the defects in the material's structure, which appear when it is formed at a high temperature. Accelerated heating of concrete followed by vibration minimizes the structural defects and so its frost resistance should be higher than after steam curing. This was confirmed completely by the studies conducted at the NIIZhB (Table 8.2). But the heating temperature affects the frost resistance of concrete also in this procedure. It can be seen quite well from the results of testing Portland blast-furnace cement concretes whose frost resistance was lower when they were heated at 95°C. The same concrete but heated at 80°C had a high frost resistance that was as good as that of concrete which hardened under normal conditions. The same values were obtained for Portland cement concretes.

An analysis of the studies gives us grounds to say that fast electric heating of concrete followed by vibration when the concrete is hot ensures high frost resistance that practically does not differ from that in concretes that hardened under normal conditions.

TABLE 8.2
Frost resistance of concretes with dense aggregates according to hardening conditions

Type of cement	Hardening conditions	Heating temp. in °C	Strength at start MPa	25		50		100		150		200	
				R	K	R	K	R	K	R	K	R	K
Low-aluminate Portland cement	Electric preheating	80	26.7	23.1	0.93	26.1	1.05	25.9	1.08	20.7	0.76	10.3	0.47
	Steam curing	80	25.2	19.9	0.83	24.4	1.02	25.4	1.11	16.9	0.65	16.3	0.58
	Normal conditions	—	25.3	19.9	0.812	24.4	1.00	26.4	1.12	20.3	0.93	22.5	0.98
Portland blast-furnace cement	Electric preheating	95	24.3	18.9	0.95	21.3	1.07	13.6	0.67	0	—	0	—
		80	25.6	25.6	1.11	25.4	1.07	24.6	1.01	18.8	0.78	17.8	0.72
	Steam curing	80	22.9	20.9	0.98	21.8	1.02	0	—	0	—	0	—
	Normal conditions	—	23.7	16.5	0.80	19.5	0.94	14.9	0.82	15.6	0.78	0	—
Rapid hardening Portland cement	Electric preheating	80	22.9	15.7	0.88	20.5	1.14	16.7	0.92	15.9	0.925	15.0	0.82
	Steam curing	80	21.2	16.9	1.06	14.3	0.89	12.0	0.76	0	—	0	—
	Normal conditions	—	24.0	21.0	0.88	24.0	1.00	23.3	0.99	26.0	1.04	24.8	1.05

Note: K is a coefficient of frost resistance defined as a ratio of the strength of concrete subjected to thermal treatment and tested for frost resistance to the strength of a similar concrete of the same age that hardened after the thermal treatment under normal conditions.

8.6 EQUIPMENT FOR ELECTRIC PREHEATING OF CONCRETE MIXTURES

Concrete mixture is best to be heated directly before it is placed in a form. An electric preheating bed is provided at the site for this purpose. It may be designed in different ways according to the volume of concrete to be heated.

When the volume is large and the concrete mixture can be placed without additional reloading, it is heated directly in the body of a dump truck (Figure 8.8). The concrete carrying truck is stopped at the electric heating point. A comb of plate electrodes with electric insulation fasteners to prevent the short-circuit of the electrodes to the body of the truck is lowered and fully immersed in the concrete by a telpher. Each electrode is connected to an electric phase and thus the current heats the concrete mixture as it is passed through it between the electrodes. A voltage of 220 or 380 V is supplied to the electrodes from the control panel to heat up the concrete faster. The heating to 60 or 80°C lasts for five to seven minutes whereupon the voltage is cut off and the electrodes can be easily taken out of the concrete when a vibrator installed on them is switched on. The truck moves to the concrete placing area and dumps the heated concrete directly into the structure without reloading. This heating procedure is convenient for installation of foundation slabs, flooring beds, and road and airfeld pavements. Once the concrete is placed and compacted by vibration the structure is covered and the concrete is cured by the thermos method.

When concrete cannot be unloaded from a truck directly into a structure, it is preheated in bins on the bed provided near the structure being erected. The bed is a boarded deck that carries the bins equipped with electrodes (Figure 8.9). After the concrete is unloaded from a vehicle into the bins it is evenly spread between the electrodes by means of a vibrator. The electrodes are well insulated from the walls and bottom of a bin. When the bins are full (two bins are usually installed side by side, their total capacity corresponding to that of the truck that brings the concrete), power cable are connected to the electrodes and a voltage of 220 or 380 V is supplied from the control panel. Hot concrete from the bins lifted by crane is unloaded into the structure, compacted, and covered.

With the above-described methods concrete is preheated in batches. A continuous electric heating unit was designed recently by a group headed by Prof. A. S. Arbeniev. It is a pipe four to six m long, where concrete is fed from one end and comes out already heated from the other end. This unit can be mobile moving along as a structure is concreted since it is convenient to feed concrete directly into the formwork from it without any additional handling. Inside the pipe are electrodes where a voltage of 220 or 380 V is supplied.

An ingenious method for electric preheating of concrete mixtures was devised by L. M. Kolchedantsev. The continuous heating unit is combined with a bin. Concrete is discharged into the bin from a mobile mixer. After it is filled up the bin is taken to the concreting site. The concrete passes from the bin through the pipe with electrodes and then is delivered to the formwork when heated to 60 or 70°C (Figure 8.10). Placed concrete is compacted and is covered immediately with vapor insulation and heat insulation for further thermos curing until it attains the required strength. This method produced good results at construction sites in the frost down

FIGURE 8.8 The truck-mounted unit for electric heating of concrete mixtures. a — general view, b — immersion of electrodes in the truck body to heat a concrete mixture.

to –20°C. It took 40 kWh of electric power to heat up 1 cu.m of concrete. In a structure with a surface modulus up to 10 it reached 40 or 50% of the design strength in six to eight hours and 70 to 100% after 24 hours. The output of this unit ranges from three to 15 m^3/hr and the required installed power is 120 to 600 kW.

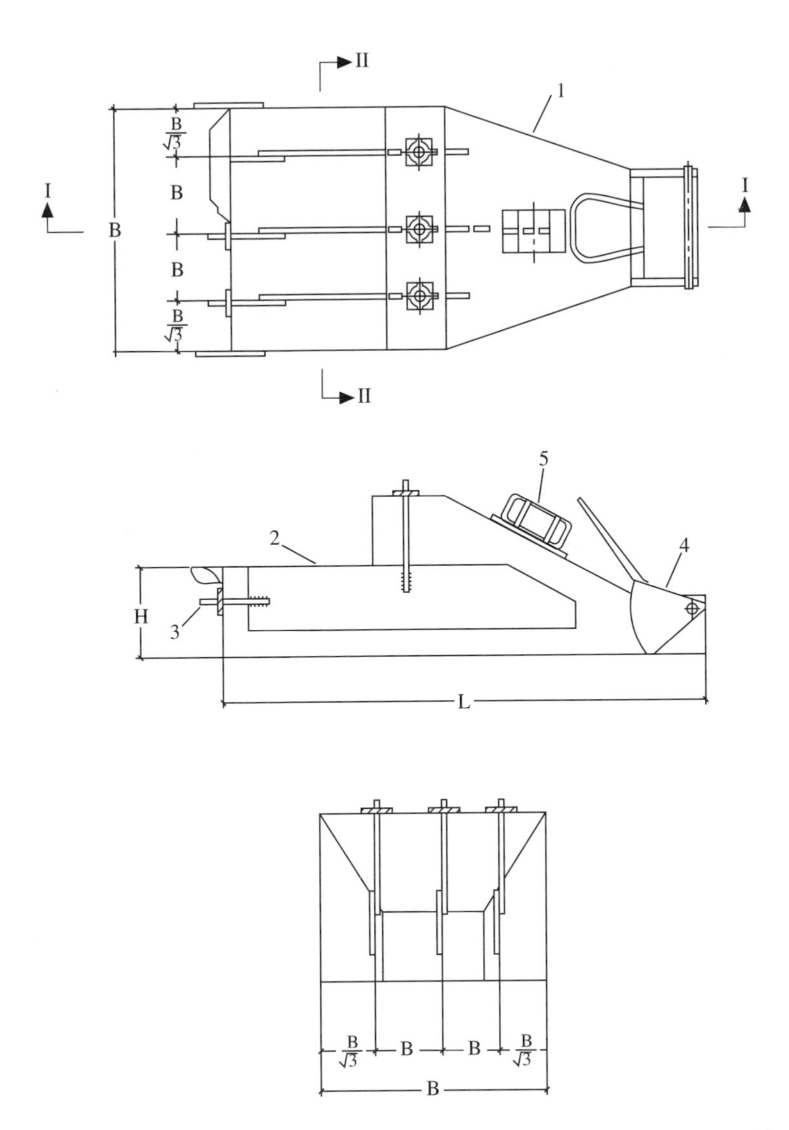

FIGURE 8.9 The schematic of a bin for electric heating of concrete mixtures. 1 — bin, 2 — plate electrode, 3 — connection of an electrode to the power cable, 4 — gate, 5 — vibrator.

The advantages of this unit are that the heat loss is minimized and the problem of the heated concrete losing its plasticity does not exist because there is no interval between preheating and placing. The concrete can be easily placed and compacted and only after that it starts setting and hardening quickly.

Special attention should be given to safety in cyclic electric heating of concrete mixtures because 220 and 380 V current is used. The beds where concrete is heated should be fenced, provided with warning lights for heating at night and with signs prohibiting people to be at the bed when voltage is on. When concrete is heated

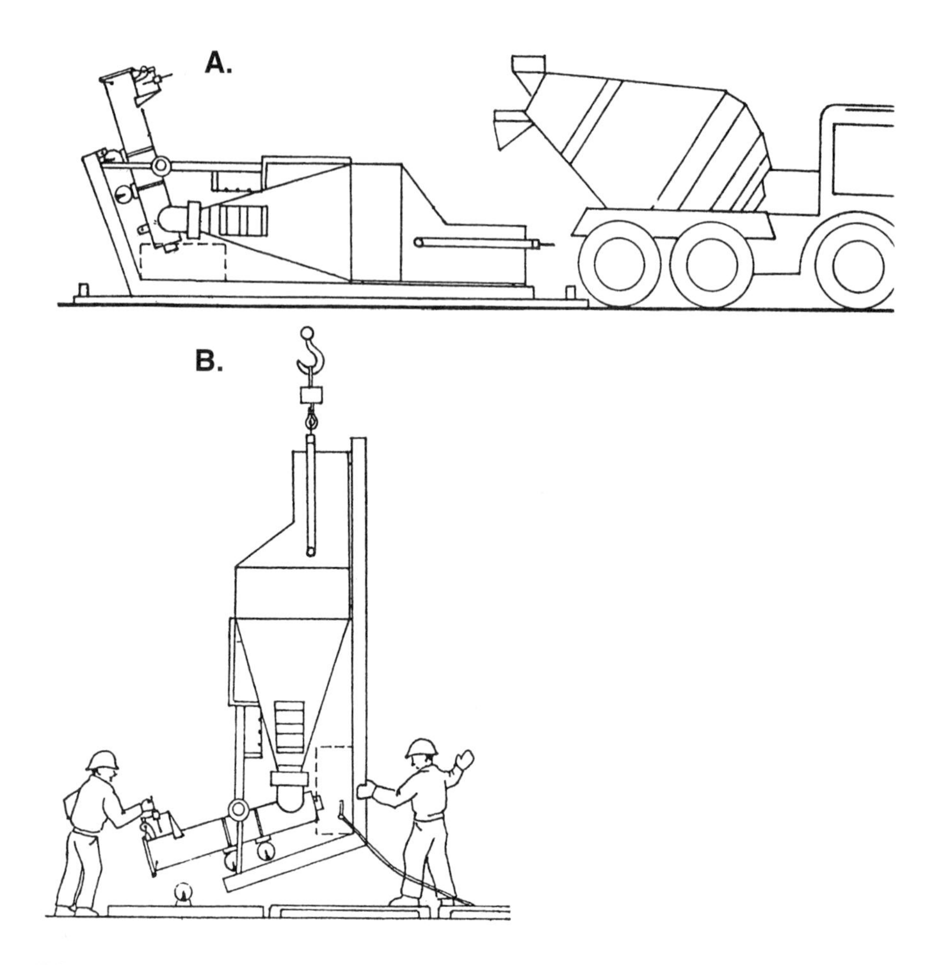

FIGURE 8.10 The schematic of a unit for electric heating of concrete mixture combined with a bin. a — position when the bin is loaded with concrete, b — position when concrete is placed.

continuously, the external pipe should be thoroughly insulated from the internal one where the electrodes are installed and the whole unit should be grounded.

Calculation of electric power for preheating of concrete mixtures. To use the electric preheating of concrete mixtures, we must know the required electric power. The specific power for heating one cu.m of concrete mixture can be found from the formula

$$P = P_1 \alpha$$

where: P = specific electric power required to heat one cu.m of concrete in kW/m^3
P_1 = electric power required to heat up the concrete in kW/m^3
α = coefficient that allows for heat loss and assumed to be 1.25 at the construction site.

The electric power required to heat up a specified volume of concrete mixture depends on the daily flow of concrete and on the specific energy consumption. It is given by

$$P_1 = \frac{QS}{n \cos\varphi K}$$

where: P_1 = required electric power in kW
Q = specific energy consumption in kWh/m^3
S = daily flow of concrete in m^3/day
n = working time during the day in hours
K = equipment utilization factor in time, which is 0.8 for operation with two bins for cyclic heating and one for continuous heating.

The specific consumption of energy required to heat up 1 m3 of concrete mixture can be found from

$$Q = c\gamma V(t_2 - t_1)$$

With a heat loss of 1.25 and a factor for conversion Kcal into kWh of 0.00118 the specific energy consumption per 1 cu.m of concrete is

$$Q_1 = c\gamma V(t_2 - t_1) \times 1.25 \times 0.00118$$

The density of concrete being 2400 kg/m^3 and heat capacity 0.25 Kcal/kg × °C,

$$Q_1 = 0.885(t_2 - t_1)$$

where: Q_1 = specific electric energy consumption for heating one m^3 of concrete in kWh/m^3
t_2 = heating temperature of concrete mixture in °C
t_1 = temperature of concrete at the start of the heating in °C
0.885 = product of density, heat capacity, the heat loss coefficient and a Kcal to kWh conversion factor.

The specific consumption of electric energy for heating one m^3 of concrete mixture depends on a number of factors, including the ambient temperature and wind, heating temperature, the design of the concrete heating device, etc. It can be roughly assumed to be 0.9 kWh per each degree of heating.

The approximate specific electric power according to the duration of heating is 625 kW/m^3 for heating during five minutes, 313 kW/m^3 for heating during 10 min, 213 kW/m^3 for heating during 15 minutes, and 156 kW/m^3 for heating during 20 minutes.

To reduce the electric power required to heat a concrete mixture, you can heat it in portions of less than one cu.m or use continuous heating.

To select the transformer power we have to proceed from electric energy needed for the maximum required quantity of concrete, the rate of temperature rise and the required temperature level, and the ambient temperature. The specific electric energy consumption is 40 kWh/m^3 on the average for the ambient temperature down to –10°C when concrete is heated to 60°C. For lower temperatures the required specific power increases to about one kWh per each degree of an increase in temperature.

When the ambient temperature is lower than –20°C and wind is strong, the consumption of electric energy can be as high as 1.5 to two kWh per 1°C. Knowing the amount of heated concrete and the heating rate we can easily determine the required electric power. To avoid excessive power requirements, it is good practice to heat no more than two cu.m of concrete at once. The parameters of electric heating of concrete mixtures can be found by calculation or in specially plotted nomograms.

Preliminary curing of a concrete mixture at a low above-zero temperature before heating is useful and this opportunity should not be lost provided, of course, the conditions at the construction site allow this. Voltage should be supplied to the electrodes before the concrete reaches 0°C to avoid its freezing at the contact with the electrodes and the resultant difficulties with heating. The temperature must not drop lower than 3 to 5°C at the site by the time the current is switched on.

The minimum time needed for electric heating is dictated mainly by two considerations, including the internal heat exchange and the requirements for electric power. It was found by experiments and by an analysis that the lower limit of the temperature rise rate should be five minutes. When this period is shorter, coarse aggregate has no time to heat and the average temperature of the concrete may drop by 5 to 20°C after the power is turned off. In addition, the reduction of the heating time by one minute below this limit increases the required power by about 40%.

The upper limit of the electric heating time is 15 minutes. When it is longer, the consumption of energy increases due to the heat loss to the environment and the setting time of concrete is shortened so that it becomes difficult to place it in the formwork. In addition, the output of the heater drops and that may adversely affect the work schedule.

The concrete heating temperature depends on the type of cement used in the mixture and its setting time in the hot state. It should not be higher than 70°C for Portland cement concretes and no more than 80°C for concrete containing Portland blast-furnace cement and the like.

Heated concrete should be placed in the formwork immediately after the current is cut off (Figure 8.11). The interval between the end of heating and placing should not exceed 15 minutes. Since concrete is not heated to more than 50 to 70°C at construction sites, it has enough workability for placing and compaction for 10 or 15 minutes and so does not need any additional water. However, in some cases when the content of C_3A in cement is high or when concrete is heated to 70°C or higher the water content may be increased by 5% in mixing without jeopardizing the quality of the concrete.

FIGURE 8.11 Placing of a preheated concrete mixture in the structure of a coal hopper erected in winter in Siberia.

The need to place hot concrete quickly is also dictated by considerations of external heat exchange because the heat loss increases considerably and its average temperature drops when the mixture is kept longer before placing.

Hot concrete should not be reloaded after it is delivered to the placing site because such operations cause high heat losses. If we compare the heating of concrete in dump trucks and in bins from this point of view, we can say that the latter method is preferable because the concrete goes from there directly into a structure without any additional handling. Concrete should be heated in trucks only where it is to be placed directly. Concreting of slabs such as road and airfield pavements and floors in industrial buildings, and of foundations in this fashion is a good solution. The reloading of concrete mixtures reduces the efficiency of the method and effective measures should be taken to reduce the heat loss if the reloading of concrete cannot be avoided. It can be achieved by insulating the bins and other means of feeding concrete into a structure.

FIGURE 8.12 Foundations for columns concreted after the concrete mixture was electrically preheated.

Heated concrete is placed and then vibrated whereupon its exposed surface is covered with a heat insulator on top of vapor insulation. Thermos curing is generally used but cooling of the most vulnerable spots, such as corners, protruding parts, and surface layers, must be closely watched. By the time it cools down to 0°C, the concrete's strength should not be lower than that required by regulations, including the Building Code. From practical experience we can say that concrete in massive structures with a surface modulus up to eight and an ambient temperature down to –40°C attains its critical strength by the time of freezing. Concrete in structures with a higher surface modulus should be covered more thoroughly when the ambient temperature is low or even heated if the cover is not enough.

Structures erected from electrically preheated concrete has a good appearance (Figure 8.12). The concrete is very dense and strong. The reason is a partial loss of water during heating and placing and self-compression of the concrete in cooling because the volume of its gaseous phase inside contracts more than solid constituents thus reducing the internal pressure.

High internal stresses occur in structures with a low surface modulus in winter due to a gradual heating of the core by the heat of hydration. Placing of hot concrete reduces the internal stresses and thus helps to solve this difficult problem to some extent.

8.7 CONCRETING PROCEDURES WITH PREHEATED CONCRETE MIXTURES

Concrete mixtures to be preheated should be sufficiently plastic (the minimum slump six to eight cm). The laboratory should provide the operator at the control panel with data on the specific resistance of the concrete. It can vary considerably due to introduction of chemical agents, particularly antifreeze electrolyte admixtures. These may reduce the specific resistance by half or more depending on the type and quantity of an admixture added. Small quantities of antifreezers are added to concrete to prevent freezing of the mixture during transportation in severe frost. Their impact on specific resistance is higher than an increase in water content. So the proportioning of water and chemical agents of any type should be accurate and deviations should not exceed ±1% of the amount of the component added.

It is good practice to have a low temperature of concrete mixtures as they are discharged from a concrete mixer or a batching plant to avoid excessive consumption of energy for heating aggregates and water. It should be borne in mind, however, that the temperature of concrete must be above zero when it is delivered to the heating site. Frozen concrete has a very high specific resistance and cannot be heated with electrodes. It was found in practice that the temperature of concrete mixtures should be at least 10 to 15°C at the outlet of a mixer. When the subzero ambient temperature is close to 0°C, concrete taken from the batching plant may be colder.

Water from heavy snowfalls or rain that occur sometimes during thaws must not get into concrete as it is carried from the batching plant. Extra water in a concrete mixture not only increases its water–cement ratio and thus reduces the strength and durability of the concrete but also changes its specific resistance, which affects the conditions of electric heating.

The concrete delivered to the construction site should be heated at once in the same vehicle or in special bins provided with electrodes. The formwork must be fully ready for concrete placing, cleaned of snow and ice, and insulated if required by design.

Heated concrete is quickly placed, compacted and its exposed surface covered with a heat insulator on top of vapor insulation. As the concrete is cured in a structure, the temperature at the spots that are apt to cool fast is constantly monitored. Should the ambient temperature drop sharply and thus make the concrete cool faster, there may be danger of some parts of the structure being frozen. An additional cover should be provided in this case and the structure may be even heated if this cover is not sufficient.

Building contractors consider the method of electric preheating quite economical, its costs being lower than those of electrode heating which is one of the least expensive electric heating procedures.

Electrically preheated concrete mixtures have been used to build millions of cubic meters of various structures in cold weather in Russia and these behave very well in service. The wealth of experience in the application of this method at winter construction sites in Russia for more than 20 years allows us to recommend it as one of highly effective and cost-saving methods of winter concreting.

Safety precautions in electric preheating of concrete mixtures. Concrete mixtures are preheated by electric current of 220 or 380 V. So care should be taken to ensure safety of the heating. Certain basic precautions make this method absolutely safe. There has not been a single case of human injury for the many years that this method was used in Russia.

Particular emphasis is placed on the equipment of the bed for electric preheating. Bins with electrodes are installed on a boarded deck and are always grounded. The places where power leads are connected to a bin should be kept clean and closed with protective caps when the bins are filled with concrete.

The electric preheating bed is fenced and warning signs are attached to the fence to prohibit entry of the personnel in the course of electric heating. The bed should be well lighted at night and red signal lamps on the fence should warn the personnel when the voltage is on.

All repairs of the bed and bins, spreading of concrete in them, and inspection of wires and of connections can be done only when power is off. It also refers to temperature measurements where thermometers are used. For this reason, remote control using thremocouples or temperature-sensitive elements to be installed in the bins before heating is started are preferable.

The personnel in charge of electric heating of concrete should be properly trained and the operation of the bed should be supervised at all times.

When concrete is heated in a dump truck or another vehicle, a particular attention is given to the fasteners made of an electric insulation material, which prevent short-circuit to the steel body of the vehicle. The electric preheating bed where trucks enter is fenced and also has warning signs and red signal lights at night

Electricians on duty should have the equipment required for measuring electric characteristics of concrete heating, rubber boots, and rubber gloves. The same refers to the laboratory workers who check the temperature of electric heating and take concrete samples.

When properly taken, elementary safety precautions can prevent any troubles or accidents in the course of electric preheating of concrete mixtures.

8.8 PREHEATING OF CONCRETE MIXTURES IN TRUCK MIXERS

The preheating procedure in casting a concrete structure in place usually involves mixing, transport of concrete to the construction site, accelerated heating near the concreting area, fast placing, compaction, and covering to prevent concrete from cooling too fast. When concrete is heated in a truck mixer the procedure changes and the heating is combined with transportation. Electricity cannot be used to heat concrete as it is transported and so a chemical method has to be adopted.

The idea of the chemical method developed under the guidance of Prof. B. M. Krasnovsky consists in adding a heat evolving chemical complex containing aluminum powder, caustic soda, and sodium pyrophosphate to concrete during mixing and transport. The procedure is as follows. Proportioned aggregates and aluminum powder are loaded in a mixer and then agitated for one or two minutes until the powder is evenly distributed over the entire volume of the aggregates. Then a solution of caustic soda and a part of mixing water (25%) are added whereupon the mixture is stirred again for one or two minutes until all the components are wetted and then it is kept in the mixer undisturbed for 12 to 15 minutes. The aggregates start to heat up as soon as the caustic soda solution and mixing water are introduced. The heating is completed in 10 to 12 minutes.

Once the aggregates heat up, proportioned cement is loaded into the mixer and the mixture is agitated for one or two minutes. Then the remaining water and a solution of pyrophosphate are added. The mixture is stirred for one or two minutes, a plasticizer solution is introduced and the mixing continues for another two or three minites. Thus, it takes 23 to 25 minutes to heat the aggregates and prepare the concrete mixture.

TABLE 8.3
Macroporosity of concrete with and without preheating

	Distribution of macropores by size (μm) in % of total strength					
Concrete specimens	**1 to 100**	**101 to 200**	**201 to 300**	**301 to 400**	**401 to 500**	**Over 500**
Concrete that hardened under normal conditions	24.0	33.5	8.4	2.7	9.8	21.6
Concrete preheated in truck mixer	40.6	35.2	7.7	5.7	7.6	3.2

The procedure of mixing concrete with the heat evolving chemical complex may be different. The aggregates can be heated to a temperature of about 70°C in a special tank rather than in the truck mixer. Then the hot aggregate is loaded in the truck mixer and cement, the remaining mixing water, and the sodium pyrophosphate solution are added. The plasticizer is added after stirring and the mixture is carried to the placing area. Thus, the mixer utilization time for mixing and transport of heated concrete does not differ much from its ordinary use for delivery of concrete mixture.

A study of the properties of concrete preheated by the chemical method found that it had a high frost resistance that exceeded that of the same concrete but not chemically heated. The reason is increased porosity of the concrete due to a great number of smallest pores (Table 8.3). However, the strength of the concrete at the age of 28 days was 15 to 17% lower although it reached the design strength on the 90th day. By increasing the cement content by 10% we can make the concrete that underwent chemical preheating attain its design strength in a month.

The preheating of concrete mixtures with the heat evolving chemical complex has been used for installation of foundations at various building projects and produced good results.

It should be noted, however, that the chemical heating of concrete during transport is more complicated than electric preheating at the construction site. For this reason it is better to be used in the cases when electric heating is impossible because of lack of electricity or for any other reason.

9 Induction Heating of Concrete

9.1 THE PRINCIPLE AND APPLICATION OF THE METHOD

The method of induction heating of concrete to accelerate its hardening was developed in Russia by a group of researchers, including Prof. B. M. Krasnovsky, V. S. Abramov, and M. Sh. Tulemyshev, headed by Prof. N. N. Danilov. The idea of the method is that concrete is heated by steel elements warmed up in an electromagnetic field.

The electromagnetic field is generated round a conductor by electric current passed through it. When placed in this field, a ferromagnetic element warms up due to eddy currents (Foucault currents) induced by the field and heats the concrete. Heat is transferred to the concrete from heated steel reinforcement, steel inserts, and steel formwork by conduction without any intermediate thermal resistances.

The electromagnetic field is generated by multiturn inductors made of a high-conductivity material. The inductors are either wound round structural members to be heated if they are of linear type (columns, beams, joists, or piles) or are placed in the form of coils in the plane of the members, such as panels or slabs.

The electromagnetic field appears to affect to some extent the hardening processes that occur in concrete. It may influence the structure and properties of water with hydration products dissolved in it, the orientation and aggregation of dispersed particles in the multicomponent material, the dispersion of hydrate formations, and the direction of the formation of the structural crystalline lattice thus making perhaps the system anisotropic. Unfortunately all this has not been proved yet to any great extent by experiments and it is hard to be certain today about the part played in reality by the electromagnetic field and about its impact on concrete let alone any practical application of the theory.

Keeping in mind the above-described state of affairs with the theory of the matter and its practical proof, we may say that heat is transferred to concrete there through contact, i.e., by conduction. So all the peculiarities of concrete structuring, structural defects, heating conditions, etc. appear to be the same as in electric contact heating.

The manner in which heat is supplied and the key role of ferromagnetic elements inside and in contact with the concrete dictate most appropriate applications of the method in construction for acceleration of concrete hardening. Concrete is best be heated in the electromagnetic field for erecting cast-in-place structures with heavy reinforcement evenly spread across the cross section. The formwork may be wooden or metallic but the latter should be less than four mm thick to avoid shielding of the electromagnetic field. When made of such steel sheets, it will act as a heater and will not hinder the heating of the reinforcement. The formwork made of thick plate will shield the electromagnetic field getting warm itself but not letting the field to heat the reinforcement.

The induction heating is a good type of thermal treatment for reinforced concrete columns, joists, beams, piles, sealed joints of precast units, and reinforced concrete structures erected in slip forms.

9.2 DETERMINING BASIC PARAMETERS OF HEATING

The intensity of heat evolution of ferromagnetic elements in reinforced concrete structures under induction heating does not depend on the electrophysical properties of concrete as, for example, in electric contact heating. It depends on the intensity of the magnetic field and on the electric and magnetic properties of the heat sources, i.e., steel reinforcement, steel formwork, and inserts. The steel elements that heat up in the electromagnetic field are called "charge."

The eddy currents that arise in the charge do not circulate over the entire thickness of the elements. The electromagnetic wave gradually damps down as it penetrates deeper into the metal. For this reason the density of electric current and intensity of the electric and magnetic field under alternating current have the highest values near the surface of the charge.

The amplitude of the damped electromagnetic wave drops sharply in the surface layer of the charge and Joule's heat generated by induced currents is 90% and equals numerically the heat which would have evolved if the current circulated only in this layer and had a constant density in it. The thickness of the layer called the "depth of current penetration into metal" is given by

$$A_s = 5030\sqrt{\frac{\rho}{\mu f}} \qquad \text{cm}$$

where: Δ_s = depth of current penetration into metal in cm
ρ = specific electric resistance of the charge material in ohm.cm
μ = relative magnetic inductive capacity of the charge material
f = frequency of electric current in Hz.

The output magnetic power from the charge surface unit (ΔP) with the specific resistance of the material ρ, relative magnetic inductive capacity μ, and the frequency f in the magnetic field of the intensity H is

$$\Delta P = \rho_n H^2 \qquad \text{W/cm}^2$$

where: ρ_n has the physical meaning of the specific surface electric resistance of the charge material.

The dependence of ΔP and ρ_n on the intensity of the magnetic field H for steels with different electric resistances can be found in the diagram shown in Figure 9.1.

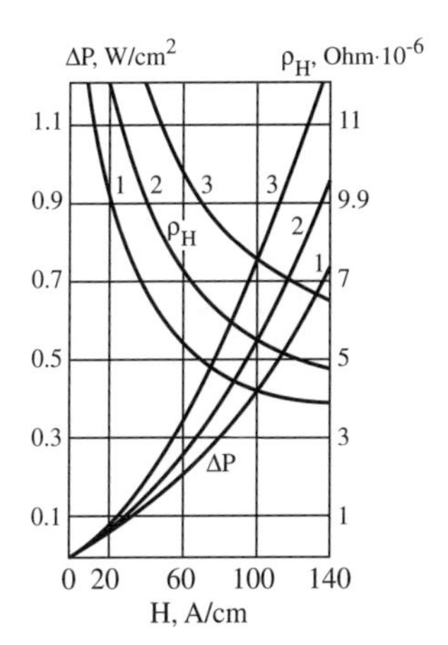

FIGURE 9.1 The relationship of ρ_n and ΔP and the intensity of the magnetic field H. 1 – $\rho_n = 10 \times 10^{-6}$, 2 – $\rho_n = 20 \times 10^{-6}$; $\rho_n = 30 \times 10^{-6}$ ohm × mm^2/m

The specific active power ΔP and other parameters of induction heating can be found on the basis of the total active power P_{as} required to ensure specified thermal treatment conditions. It is made up by the electric power needed to heat the concrete up, electric power needed to heat the formwork, and power to compensate the heat loss to the environment in the course of concrete heating taking into account the power equivalent to the heat of hydration. To take off the power required to heat concrete under the specified conditions from the charge surface unit we must find the number of turns of the inductor. For the given voltage this number of turns should ensure the appropriate intensity of the magnetic field, under which the required power from the charge surface unit is provided.

The temperature parameters of heating, including the maximum heating temperature and the rate of temperature rise, are set in advance. These parameters do not differ from those required for any concrete heating method (except preheating). The maximum heating temperature usually lies between 50 and 90°C for any concrete according to the type of cement. It is 70 or 80°C for Portland cement concretes. The rate at which temperature rises on the surface of the charge depends on the massiveness of a structure. It is 10 to 15°C per hour for structures with $M > 10$ and 5 to 8°C per hour for massive structures with $M < 6$.

Once the parameters of the heating conditions are set the electric power required for the heating-up period and for isothermal heating is determined.

The next stage in the calculation is to find the area of the charge surface, which is in contact with and transfers heat to concrete. For bar-type reinforcement we should find the total area of the reinforcement elements assuming that all the charge elements positioned in parallel to the axis of the inductor in the cross section of the structure to be heated are equivalent because the intensity of the electromagnetic field varies but little over the cross section of the structure:

$$F_s = \pi dhn$$

where: F_s = total area of the charge when the structure is reinforced with bars in a nonmetallic formwork in cm^2
d = diameter of reinforcing bars in cm
h = bar length in the inductor action zone
n = the number of rebars in the cross section of the structure.

For structures with the bar-type reinforcement cast in a metallic formwork the charge surface area is

$$F_s = \pi dhn + F' + F''$$

where F' and F'' are areas of the outer and inner surface of the formwork in cm^2, respectively.

It should be emphasized that this formula is suitable only for the case when the steel formwork does not shield the electromagnetic field. Otherwise only the formwork becomes a heat generating element.

On the basis of the above data we can find the specific heat power of the charge, which is required to heat the structure, as follows:

$$\Delta P_p = \frac{P_p}{F_s} \text{ and } \Delta P_1 = \frac{P_1}{F_1}$$

where ΔP_p and ΔP_1 is the required specific heat power to heat concrete at the heating-up stage and at the stage of isothermal heating, respectively, in W/cm^2.

Then we find the total resistance of the inductor-charge system:

$$Z = \sqrt{r^2 + (\omega L)^2}$$

where: Z = total resistance in ohm
r = active resistance of the inductor-charge system, which is $r = r_i + r_s$ ohm
ωL = inductive impedance of the system, which is

$$\omega L = \omega L_i + \omega L_s \text{ ohm}$$

where: r_i and ωL_i = active resistance and inductive impedance of the inductor
r_s and ωL_s = active resistance and inductive impedance of the charge

The inductor for induction heating may be cylindrical or rectangular.

The active resistance of the cylindrical inductor can be found as

$$r_i = AR_i$$

where: $A = 1.26 \times 10^{-5}$ ohm/cm for the copper wire of the inductor

$A = 1.66 \times 10^{-5}$ ohm/cm for the aluminum wire of the inductor
R_i = radius of the cylindrical inductor in cm, which is

$$R_i = R_k + \delta$$

where: R_k = radius of the structure in cm.

The active resistance of the rectangular inductor can be found as

$$r_i = \frac{A}{\pi}(a_i + b_i)$$

where: a_i and b_i = lengths of the cross section sides of the rectangular inductor in cm:

$$a_i = a_k + \delta$$

$$b_i = b_k + \delta$$

where: a_k and b_k = lengths of the cross section sides of the structure in cm
δ = thickness of the formwork

The inductive impedance of the cylindrical inductor is

$$\omega L_i = BR_i\beta$$

where: $B = 1.24 \times 10^{-5}$ ohm/cm^2
β = dimensionless shape factor of the inductor to be found in a diagram (Figure 9.2) according to the ratio of the inductor length to its radius

$$\frac{h}{R} \text{ or } \frac{h\pi}{a_i + b_i}$$

The inductive impedance of the rectangular inductor is

$$\omega L_i = \frac{B}{\pi} a_i b_i \beta$$

The active resistance of the charge is found as

$$r_s = \Pi_s \rho_n$$

where: Πs = sum of parameters of the charge in cm

For reinforcing bars of a structure in a nonmetallic formwork

$$\Pi_s = \pi dn \text{ cm}$$

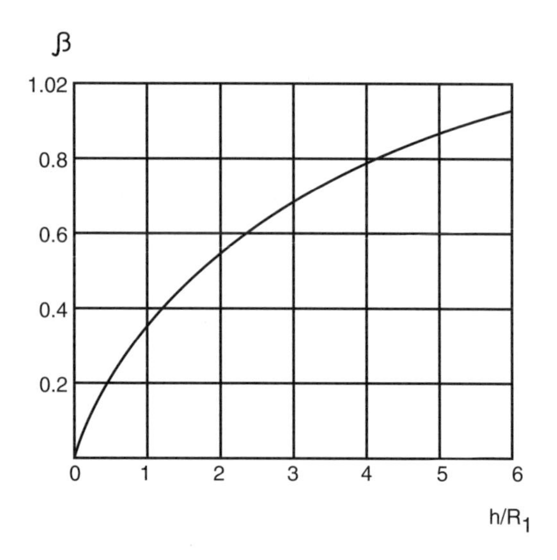

FIGURE 9.2 The dependence of the inductor shape factor β on the ratio h/R_1.

For reinforcing bars of a structure in a metallic formwork

$$\Pi_s = \pi dn + \frac{F''}{h} + \frac{F'}{h}$$

Since the intensity of the magnetic field H during the heating-up period differs from that for isothermal heating, their active resistances and inductive impedances are also different. So the total resistance of the inductor-charge system in each period is different. Having calculated the total resistance during heating, Z_1, and during isothermal heating, Z_2, and assuming a certain voltage value chosen for induction heating we can find the number of inductor turns for heating of placed concrete under preset conditions as follows:

$$N = \frac{V_1}{Z_1 H_1}$$

where: N = the number of inductor turns
V_1 = selected electric voltage in V
Z_1 = total resistance of the system during the period of heating in ohm
H_1 = intensity of the magnetic field in the heating-up period

The strength of current in the inductor for the established number of turns is found from

$$I = \frac{H_1 h}{N} \text{ A}$$

The cross section of the inductor is selected according to the strength of current. The resultant strength of current may prove to be unacceptable and so it is recalculated. For the given permissible strength of current I in specific conditions and maintaining the intensity of the magnetic field constant we find again the number of inductor turns as follows:

$$N = \frac{H_1 h}{I_1}$$

From the number of inductor turns thus obtained we determine the required voltage

$$V_1 = NZ_1$$

and the intensity of the magnetic field

$$H_1 = \frac{H_1^2 Z_1 h}{I_1} V$$

For the period of isothermal heating the calculation of the parameters of the induction heating method is reduced to finding the voltage to be supplied to the inductor with the established number of turns. The intensity of the magnetic field created then ensures the output power from the charge, ΔP_1 .

$$V_2 = NZ_2H_2$$

where: V_2 = voltage supplied to the inductor during isothermal heating in V
Z_2 = total resistance of the system during isothermal heating in ohm
H_2 = intensity of the magnetic field during isothermal heating in A/cm.

Certain specific features of induction heating should be taken into account when it is organized. Thus, according to the theory of induction heating, the intensity of the magnetic field decreases in inductors of finite length along the inductor from the center to the edges and across the cross section from the surface to the center. So the heat power, which is proportional to the intensity of the magnetic field near the edges of the inductor is lower in the center of the cross section than in the middle of the inductor height on its surface. This causes temperature differences across the section and along the height (or length) of the structure being heated if it is reinforced uniformly over its entire height (or length if the inductor turns are distributed evenly).

However, it was found in practice that the intensity of the magnetic field varies little across the cross section of the inductor and its impact on the temperature field may be neglected. The intensity of the magnetic field varies very much over the height of the inductor decreasing at the edges nearly by half as compared with the center. It may lead to a considerable nonuniformity of heating over the height (length) of a structure, which may be even greater because of high heat loss to the environment at the end faces of the structure being heated.

The active resistance of the unit area of a part of the charge increases due to the lower intensity of the magnetic field near the end faces. It results in a higher total resistance of the charge, lower current strength for the specified voltage in the inductor, and thus in an additional decrease in the intensity of the magnetic field over the entire length of the inductor. Hence, the conclusion: uniform distribution of the inductor turns over the height (length) of the structure to be heated creates a nonuniform magnetic field along the inductor's length with a lower intensity, and so does not ensure the specified heating conditions.

This can be avoided by installing the inductor with different pitch at different sections of the height (length) of the structure. Thus, the turns have an equal pitch in its central part equal to 3/5 of its height (length). In extreme sections the distance between the inductor turns gradually decreases to the edges and becomes twice as short near the end faces than in the center. Zonal calculations can be used to find exactly the distribution of turns near the end faces of a structure.

9.3 INDUCTION HEATING OF CONCRETE IN FRAMED STRUCTURES

Framed structures are the most suitable for induction heating of concrete in them. The best effect is produced by heating in a steel formwork as compared with a wooden one because more heat is released for the same current in the inductor. So a lower strength of current is needed for heating concrete in the steel formwork than in the wooden one and hence lower installed power. However, the erection of structures in a steel formwork causes a large heat loss to the environment, which must be compensated. The loss can be reduced by insulating the formwork.

The induction heating of concrete follows the same rules as other thermal treatment methods. Namely, the rate of the temperature rise should be as slow as possible but not exceed 20°C per hour. The temperature of isothermal heating depends on the type of cement used and varies over a wide range from 40 or 50°C for concretes with alumina cement to 90°C for concretes containing Portland blast-furnace cement or pozzolana cement. When the heating is completed, the concrete should also cool down as slowly as possible at a rate of 6 to 15°C per hour depending on the massivenes of a structure, the forms to be struck when the temperature difference between the environment and the concrete does not exceed 20°C. The induction heating unit should be designed on the basis of the adopted thermal treatment conditions.

A well-organized induction heating of concrete depends on how carefully it is prepared. The inductor is designed and templates with grooves spaced exactly as the inductor turns are made before the start of heating, or rather before concreting. The templates, which are actually boards with grooves, are installed on the formwork and then the inductor is placed, its turns being fixed in the grooves. The latter may be replaced by nails. The inductor is an insulated wire of a calculated cross section, which is connected to power leads.

When an inductor is installed, for example, to heat columns, the high heat loss at the ends should be taken into account. So it is advisable that the inductor should be positioned above the top of a column by about 20 cm with its turns closely spaced four to six cm between them.

The reinforcement is warmed up to an above-zero temperature before concreting if necessary. It is done only in the cases when there is ice on steel. It should be remembered that the heating of the reinforcement requires plenty of energy since a lot of heat is lost from reinforcing bars to the cold environment and so they heat up slowly. It is better, therefore, to heat up even thick bars directly in concrete after the latter is placed. All the heat released in this case by the reinforcement is transferred to the concrete and steel warms up in minutes.

Induction heating starts after the placing and compaction of concrete are completed. If the heating rate exceeds the permissible value for some reason, concrete may be heated by switching the inductor on and off until the temperature of isothermal heating is reached. The same method can be adopted for maintaining the isothermal heating temperature in concrete if it becomes hotter than specified when the inductor is constantly on.

The voltage for the induction heating method usually does not exceed 120 V. But it may be higher — 220 or even 380 V — provided the electric insulation of the inductor is good enough.

An advantage of the induction heating is that concrete can be heated to any strength value since heat evolution depends only on the physico-technical characteristics of the reinforcing steel and formwork and does not depend at all on concrete as it happens with electric contact heating. When the required thermal treatment conditions are maintained, the reinforced concrete structures heated this way are as good in quality as the ones erected in summer.

The induction heating has been used to advantage at many construction projects in Moscow (Figure 9.3). The quality of the structures where concrete was heated by this method met all the specifications. But it still seldom used as compared with other electric heating methods. The reasons are higher energy consumption per one m^3 of heated concrete than with other methods of electric thermal treatment and the complexity and difficulty of installing the inductor. The latter is made separately for each particular structure. All the attempts to design a universal inductor for reinforced concrete structures of different cross sections have failed. It also proved to be quite difficult to develop a telescopic inductor for structures of the same type, which could reduce dramatically the labor input required for installation and dismantling.

Induction heating requires certain safety precautions during operation. The main thing to focus attention on is the reliability of electric insulation of the inductor and power leads, of its connections to the power leads and all the wiring and electric connections in general. All other requirements are the same as for other methods used for electric thermal treatment of concrete.

FIGURE 9.3 An inductor installed in a wooden formwork for heating concrete in a column base.

10 Sealing Joints between Precast Units and between Precast and Cast-In-Place Members in Composite Construction

Buildings and structures continue to be erected on a large scale in winter as well as in summer. Joints between precast elements are an important problem in this type of construction in cold weather. Joints are sealed with cement concretes or mortars. The sealing material may perform different functions. It may connect precast reinforced concrete units into a whole, i.e., sealing concrete may perform a load-bearing function or it may become monolithic with steel connections (bolts or welds) protecting them against corrosion and making an appearance of a solid reinforced concrete structure.

The volume of sealing concrete is usually small and to make it harden at a subzero temperature under these conditions is a difficult problem.

10.1 SPECIFIC FEATURES OF JOINTS BETWEEN PRECAST CONCRETE UNITS

The sealing material may function in different ways according to the type of joint. In load-bearing joints, sealing concrete fully participates in the work of a structure withstanding all the loads that may be transferred from one reinforced concrete element to another. Concrete in this type of joints should have a strength no lower than that of the elements to be joined, should adhere very well to the concrete of the elements, and protect securely the reinforcement in the joints against corrosion. In some joints, e.g., in structures designed for storage of liquids, such as water or industrial effluents, sealing concrete should have in addition high density and impermeability.

In joints of load-bearing or non-load-bearing structures, where reinforced concrete elements are connected by welded steel inserts, the sealing concrete fills the space of a joint and protects the steel elements against corrosion during service. Concrete in such joints should have a sufficient density and bond well with the joined reinforced concrete elements and with the connecting steel. In joints without steel parts, e.g., longitudinal joints of floor slabs, the sealing concrete is a filling and may even freeze after it attained the critical strength allowed for this grade of concrete.

In joints between structural members that separate environments, i.e., external wall panels, concrete fills a part of a joint. It should have a good bond with the concrete of the joined elements and a high density. The impermeability of the joint to water and heat insulation are ensured by construction techniques and by other materials specifically designed for this purpose.

A great number of various types of joints between precast reinforced concrete units have been inspected and defects found there provided grounds for developing reliable methods of joint sealing in winter.

A very common defect is the absence of bond of sealing concrete with the concrete and inserts of joined elements. A crack appears at the interface and thus makes a load-bearing joint fail because its seal does not transfer loads from one element to another. The crack in a load-bearing joint can also cause corrosion of reinforcement connections or inserts. The reasons for the appearance of cracks at the interface of sealing concrete with a joined reinforced concrete element are:

- Shrinkage in sealing concrete.
- Poor preparation of the surface to be joined for the sealing.
- Freezing of the concrete.
- Poor compaction of sealing concrete.

Cracks in joints. Concrete placed in a joint between reinforced concrete elements may shrink. As it contracts in volume, sealing concrete breaks away from the concrete of one of the elements and a crack appears at the interface as a result. The best way to avoid cracking at the interface because of shrinkage is to use expanding or nonshrinking cements in sealing concretes. Concrete containing such a cement expands in hardening and fills the joint tightly. Joints between precast reinforced concrete elements of structures for storage of liquids are practically impossible to seal well without concrete with expanding cement.

Another common reason for the poor adhesion of sealing concrete to the surface of a precast element is the presence of a weak cement or ice film at their interface. The cement film on the surface of a precast element appears mostly due to an intensive mass exchange during hardening, which dries the surface layer of the concrete, loosens it a little, and results in a lower hydration. Naturally, the adhesion of concrete placed in the joint is very poor. The appearance of a crack at the interface with such a surface of a reinforced concrete element can be avoided by cleaning it thoroughly of the weak film and making the surface rough. It is usually done mechanically.

The ice film on the surface is often not removed in the hope that it will melt after warm concrete is placed and then heated. The film may not melt, although it is a rare occasion, depending on the temperature of the concrete in the elements to be joined, the thickness of the ice film, and on the volume of the concrete to be placed in the joint and on its temperature. But even if the film melts, the sealing concrete at the interface gets wetted and so its strength will be lower after hardening. And the rupture always occurs across the material whose strength is lower. Cracking can be avoided in this case only by cleaning carefully the space of the joint and the surfaces of the elements of snow and ice. The best way to do it is to blow hot air (but not steam) directly before concrete is placed.

The quality of sealing concrete at the interface with the surface of a reinforced concrete element may deteriorate sometimes due to freezing. When the temperature of the ambient air and of concrete in the elements to be joined is low and the joint is narrow, sealing concrete may freeze where it contacts the elements. It will thaw later on when the concrete is heated but its strength will be lower because of structural defects. It may later produce a crack at the interface. This can be avoided either by heating the parts of the elements, which are joined, to an above-zero temperature or by adding an antifreeze admixture to the sealing concrete to prevent freezing before heating.

Poor compaction of concrete is often the result of incorrect design of its composition. Joints that have protruding reinforcement and inserts have a lot of steel in them. These joints or joints with narrow spaces cannot be filled with a low-slump or stiff mix. It is very hard to compact, which is the cause of the appearance of cavities and voids that weaken considerably the bond of the sealing material with the surface of the joined elements. It can be avoided by using high-slump mixes made plastic by a superplasticizer but not by additional water.

All the above-described cases should be kept in mind in sealing joints between precast reinforced concrete elements in winter.

10.2 ERECTION OF PRECAST CONCRETE UNITS

The sealing of joints is preceded by erection of precast units and the quality of the joints depends to a certain extent on the quality of the erection. It was observed in practice that a horizontal joint between panels or large blocks can be of good quality if these are installed on a plastic mortar evenly spread over the whole area of the lower element. The mortar is well compressed in this case and fills the joint uniformly over the entire area. When the ambient temperature and thus the temperature of the joined elements is low (–20°C or lower), mortar freezes in the joints up to 50 mm wide in five to seven minutes after placing. The precast element should be installed before the mortar freezes. Otherwise it will not be compressed and its quality will be low. When the mortar melts, it may deform and thus the precast element it supports may settle down. This will cause cracking in the vertical joints of the adjacent elements.

The freezing of the mortar at low temperatures before a precast element is placed on it can be retarded by adding an antifreeze admixture of sodium nitrite ($NaNO_2$) or potassium carbonate (K_2CO_3) to it in mixing. The mortar containing an antifreeze admixture freezes much slower at the same temperature (two or three times) thus enabling the builders to place the precast element peacefully and carefully as designed.

The cleanness of the surface to be joined affects to a great extent the quality of joints. Elements with dirty or iced surface must not be erected without thorough cleaning. It can be done by a cold or hot method before an element is installed. It is very difficult and labor consuming to clean the space of a joint of ice and dirt when the element is in place. Cement mortar or concrete must not be placed in a joint that was not cleaned regardless of the type of the connection.

Joints may be different and sealed in different ways according to the type of reinforced concrete units being erected. Thus vertical joints between external wall panels often have a complicated design and in addition to cement they may have chemical sealing compounds, ventilation air holes, etc. But horizontal joints between external wall panels and joints between floor slabs are generally filled with cement mortar.

Mortar with an antifreeze admixture combined with a superplasticizer are recommended for this purpose in winter. The admixtures prolong the time during which the sealing material can be handled before it freezes, ensure its high plastic properties, and minimize water content. Nonshrinking expanding Portland cements or rapid hardening cements are used as binders. Antifreeze admixtures should not cause corrosion of the reinforcement in reinforced joints and sodium nitrite is the most convenient admixture for mortars to seal such joints. When placed in a joint, mortar or concrete should be well compacted with a vibrator having various attachments or by rodding.

10.3 CURING OF SEALING CONCRETE IN JOINTS

Curing of a sealing material until it attains the required strength is the most important operational stage. Many difficulties arise in the process because the volume of mortar or concrete to be cured in the seal is small, heat loss to the environment is very high, and the building construction schedule requires fast hardening. Just like in other types of concreting two methods are used — antifreeze admixtures or heating. The pace of construction has to be sacrificed in the former case because hardening is slow and the concrete or mortar reaches the required strength only after one to three months depending on the ambient temperature and the temperature of the joined elements. Heating ensures fast strength development in concrete and the sealing material reaches the required strength (70 or 80% of the design value) in one or two days.

Concrete can be heated in joints by means of electrodes, electric heaters, or by induction.

The electrode method is convenient for heating concrete in joints without reinforcement. Joints between a column and the foundation made in the form of a shell (Figure 10.1) and vertical and horizontal joints between panels (Figure 10.2) are the most suitable for this method. Special care must be taken to prevent freezing of sealing mortar before electric heating starts. As was mentioned above, frozen mortar is a poor conductor of electricity and electric heating may become impossible. To avoid it, an antifreeze admixture must be added to the mortar in all cases not to let it freeze too fast during placing and, in addition, to increase its electric conductivity. Electric heating should be gradual as usual until the mortar attains the required strength (usually 50 to 70% of the design value).

Contact heating of joints using electric heaters has many different arrangements which can be divided into heating with heaters placed inside mortar and heating with heaters on its surface. The internal heating is the most economical as far as

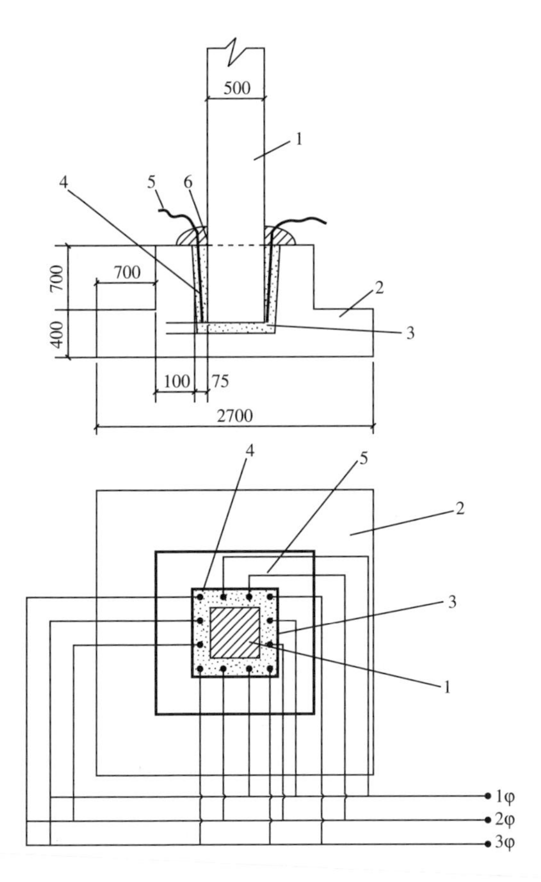

FIGURE 10.1 Electrode heating of concrete in the joint of a column with the foundation. 1 — column, 2 foundation, 3 — sealing concrete, 4 — electrodes, 5 — power lead, 6 — heat insulator.

energy consumption is concerned and it is usually done with a heating wire (Figure 10.3) that is placed in a joint before mortar is placed or embedded into it after placing if the design of the joint allows it. The heating wire is convenient for heating any joint, including a reinforced one where the heating wire is attached to the reinforcement. The exposed part of the sealing mortar must be always covered with a heat insulator on top of vapor insulation. The temperature in the joint is raised slowly.

External heating of joints is better be carried out in a heating form (e.g., joints of columns) or with external electric heaters (Figure 10.4). High temperature heaters are not suitable for joint heating because a lot of radiated energy is lost in the environment. Regardless of the temperature on the heater's radiating surface which is usually surrounded by a reflector the temperature on the concrete that absorbs the heat should not exceed 80°C. For this condition to be met, the heater is placed at a certain distance from the joint, which may be 1 m or more. So a considerable amount of radiated heat is lost in the environment because of air convection and removal of

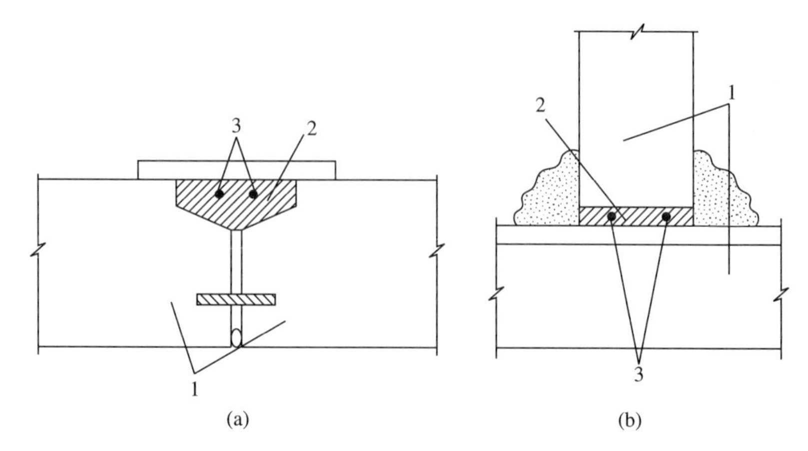

FIGURE 10.2 Electrode heating of concrete in joints between precast reinforced concrete wall panels. a — vertical joint between exterior wall panels, b — horizontal joint between interior wall panels, 1 — joined units, 2 — sealing concrete (mortar), 3 — electrodes.

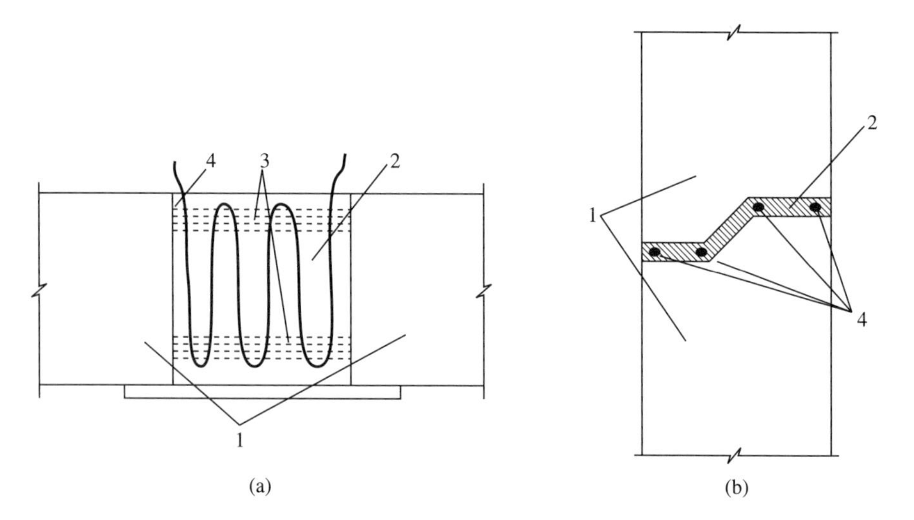

FIGURE 10.3 Heating of concrete (mortar) in joints with heating wires. a — joint of beams, b — horizontal joint between exterior wall panels, 1 — joined units, 2 — sealing concrete (mortar), 3 — reinforcement, 4 — heating wire.

heat from the joint's surface by wind. Considering the above, low-temperature heaters where the temperature on the surface of a radiator does not exceed 100°C are better be used for heating concrete in joints.

Low-temperature heaters are made from the heating wire, insulated Nichrome coils, from carbon fabric or filaments, or from any other material. These heaters are small, light, and are easy to be assembled into rigid (in the form of panels) or flexible (in the form of blankets) heating devices.

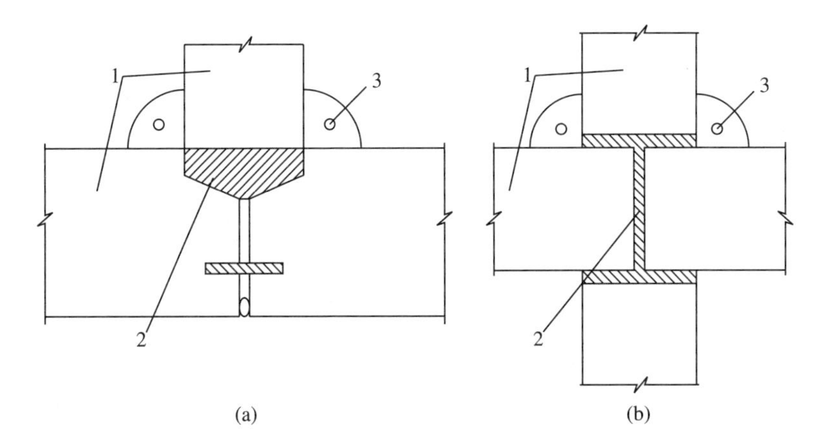

FIGURE 10.4 Heating of concrete (mortar) in joints with electric heaters. a — vertical joint of interior wall panel with exterior wall, b — horizontal joint between interior wall and floor, 1 — joined units, 2 — sealing concrete (mortar), 3 — heaters with reflectors.

Tubular low-temperature heaters are convenient for heating the space of a joint, i.e., the surfaces of the elements to be joined before concrete is placed. When the space is to be preheated, it should be tightly covered to avoid a high heat loss to the environment. The heating usually last for one to two hours until the temperature on the surfaces of the element reaches about 20°C. As soon the space become warm enough it should be filled with concrete and then usual measures taken to ensure its proper hardening.

The heaters designed for joint heating may be of reflecting type, come as heating formworks, or flexible heating blankets. A reflecting or a resistance heater may be an open coil wound round a dielectric, a tube, or a rod with a reflector that may be many shapes. The reflecting heater is installed along the entire joint and sealing concrete and the parts of the reinforced concrete units being joined are heated as specified and to a specified temperature. These heaters are easy to control since heat supply can be varied by varying electric voltage or by turning the heaters on and off according to the required heating conditions and the ambient temperature.

Heating formworks may have a steel deck or a deck made of water-resistant plywood. In any case the temperature of the heating surface of the deck should not exceed 80°C. The heaters embedded in the formwork have a reliable electric insulation. A heat insulation layer is placed on the external side of the heating formwork to prevent the heat loss to the environment.

The induction method is preferable for heating concrete in joints between linear reinforced concrete structural members such as columns or beams. A form is installed on a joint overlapping the elements be joined 30 to 40 cm on either side. An inductor is wound round the form and voltage is supplied to the inductor before concrete is placed in the joint (Figure 10.5). The protruding steel reinforcement in the electromagnetic field of the inductor and the reinforcing steel of the elements to be joined are heated. The reinforcement heats the concrete, which creates favorable conditions for filling the joint space with sealing concrete. The inductor is switched off while

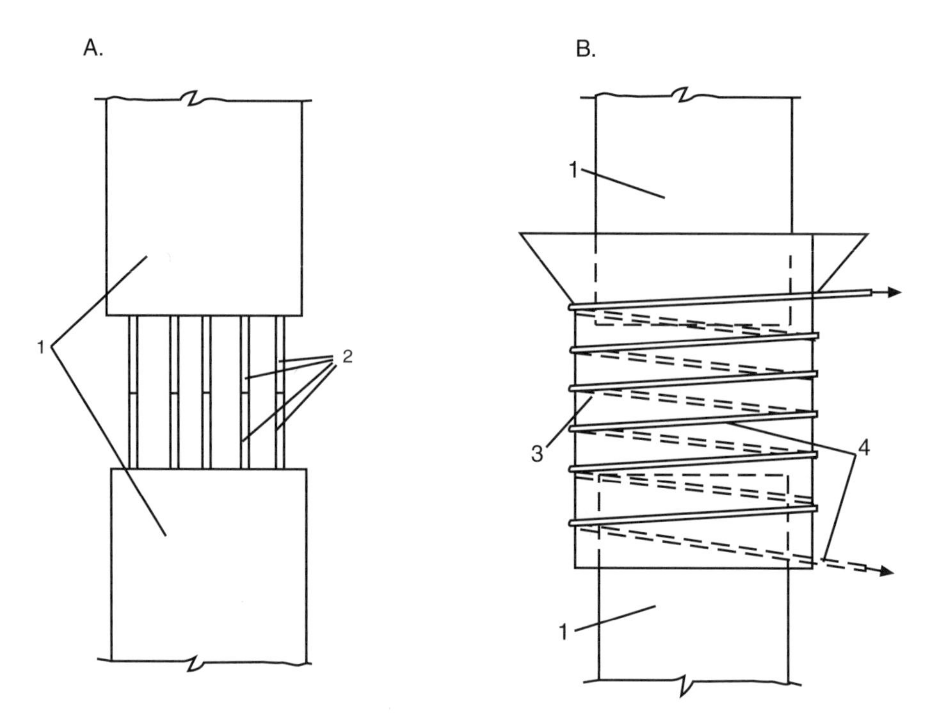

FIGURE 10.5 Induction heating of concrete in the column joint. a — joint, b — inductor installed on formwork, 1 — joined units, 2 — protruding reinforcement, 3 — formwork, 4 — inductor.

the concrete is placed and then is switched on again after the placing is completed. The heating is gradual as usual with the temperature being raised slowly until it reaches 60 to 80°C with the subsequent slow cooling and form removal when the temperature difference between the heated sealing concrete and the ambient air becomes no lower than 20°C.

Flexible heating blankets are convenient for some types of joints. A joint can be covered with such a blanket easily and tightly and concrete can be well heated as the reinforced concrete units are joined. The heating blankets are always used with a heat insulation cover. It may be either part of the blanket's design or the heat insulator is placed on top of it.

The above-described methods used to ensure the hardening of concrete in joints between precast reinforced concrete units refer to the joints where the concrete works as part of the structure or it must protect securely steel connections and the protruding reinforcement against corrosion. High demands are placed on this sealing concrete for strength and density. But there are joints where concrete is a filler and does not have to meet particularly stringent requirements. Concrete in such joints does not have time to develop any strength or sometimes even set.

Concrete is very strong when frozen but its strength drops after thawing almost to zero. Then the concrete starts to harden at above-zero temperatures and its strength gradually increases. This concrete will have structural defects, of course, including lower strength than its grade and lower density. But it does not make any difference for some joints. These include joints between blocks and panels of interior walls, longitudinal joints between floor slabs, etc. To avoid fast freezing and hanging in a narrow vertical joint during placing, 3 to 4% of sodium nitrite are usually added to mortar (cement mortar is used for such joints).

The above-listed methods for sealing joints between precast reinforced concrete units ensure fast erection of fully prefabricated buildings in winter conditions with their quality being similar to that achieved in the summertime.

11 Curing of Cast-In-Place Concrete in Enclosures

Enclosures have long been used in construction. They were widely adopted at the turn of the century for erecting structures in cold weather. Sometimes an enclosure was a real edifice of a complex engineering design, which loomed over a building being constructed. It provided quite comfortable conditions for construction in general and concreting in particular but the cost of such enclosures was so high that builders preferred to stop construction in winter. The seasonal construction was very undesirable for northern regions because cold weather lasts there for a long time. It gave an impetus to developing the construction methods that woould be cost effective considering the need to complete buildings in the shortest time possible.

Fast development of electric heating and of other methods of heating placed concrete moved the enclosures to the background and they were used rather seldom. New opportunities for them opened up with the appearance of new polymer and other effective heat insulation materials. Builders started to make enclosures of light elements that were easy to assemble and had very good heat-shielding properties. The enclosures were not so cumbersome and expensive as they used to be and they were installed only at a specific work area, i.e., they became local and did not cover completely the whole structure.

Enclosures are used now in many countries for concreting in winter and are simply irreplaceable in some cases. These include reinforced concrete smoke stacks, cooling towers, cores of high rises, silos, TV towers, and other similar structures. When tall structures are erected, enclosures not only create proper conditions for curing concrete but also provide a comfortable environment for workers. Wind always blows at great heights increasing considerably the severity of weather at a subzero air temperature and frost is felt much stronger. An enclosure creates safer conditions for the personnel, including their psychological response to height since it acts as an enclosing wall.

11.1 THE PRINCIPLE OF THE METHOD AND TYPES OF ENCLOSURES

In winter concreting, the enclosure is designed first of all for creating favorable conditions for curing concrete placed in a cast-in-place structure. The combination of this purpose with other above-described functions does not make, however, the concrete harden faster without additional measures. An advantage of enclosures is that the formwork and reinforcement can be kept warm to minimize the heat loss of concrete in placing as compared with concreting outdoors.

The diversity of cast-in-place members, elements, and structures was the reason for the development of a great variety of enclosures that differ in functional features, in space-planning arrangement, in design, and in methods of moving. By their functional features, enclosures may be divided into those that are installed only to create favorable temperature and moisture conditions for hardening of concrete and into enclosures that are designed both for curing placed concrete and for workers to warm up. Enclosures may by unheated but they protect against wind and their task is to provide better comfort for workers.

By their space-planning arrangement, enclosures may be spatial structures that cover the whole construction site or a part of it. Such enclosures have air locks and large openings with covers for delivery of materials. The spatial enclosures are made from flexible inflated shells or lightweight rigid elements that can be quickly assembled. The climate outside practically does not affect construction operations. These enclosures need a lot of material and are labor consuming to install.

Local enclosures are designed for curing concrete in cast-in-place floor slabs and other similar structural members. They are installed as soon as concrete is placed and may be insulated flat boxes or made of film, rubberized cloth, or other material stretched over a light frame. Such an enclosure is usually heated by blowing warm air inside from an air heater. These enclosures are easy to make but cold air still infiltrates in them.

Enclosures are moved from place to place in different ways depending on their type. They may be moved by crane, by jacks (when a structure is erected in sliding forms), and they may be dismantled and assembled again for concreting another structure.

The simplest are tarpaulin enclosures for concreting some structures of the "zero" cycle (substracture and utility comminications construction). They may be single-layer or double-layer types, sometimes additionally insulated. These enclosures are simple, compact, and are easy to install and dismantle.

Enclosures have been widely adopted for winter concreting in various countries, including Finland, Canada, northern regions of the U.S.A., Poland, and Denmark. By their functional features they can be divided into temporary enclosures designed for keeping an above-zero temperature in a concreted structure and into enclosures that protect against atmospheric precipitation.

Enclosure use plastics or polyethylene film 1.5 mm thick. Multilayer film covers are used sometimes, including, for example, layers of building paper or air intelayers between layers of plastic.

An analysis of the use of enclosures made of plastics has shown that the temperature under the cover can be maintained within 10 to 16°C when they are heated with portable units. The temperature in an enclosure may be as high as 16°C on sunny days without heating. It should be emphasized that all the the above refers to ambient temperatures between –5 and 10°C. When it is colder than that, a relatively high temperature can be maintained only by a more intensive heating.

Wind velocity is an important factor. The temperature in an enclosure drops noticeably as it increases, which also requires a more intensive heating. To make the temperature uniform over the height of an enclosure, we must create an artificial air convection, for example, by fans. If you do not do it, the temperature in a heated enclosure may be 20°C or more in the upper part but not exceed 4 or 5°C at the floor level. Naturally this nonuniform temperature inside the enclosure is extremely undesirable both for the personnel and for concrete hardening.

11.2 CURING OF CONCRETE IN ENCLOSURES

Enclosures that are made over structures to be concreted and which do not have any heaters are called "unheated enclosures." Their main task is to protect the work area from bad weather, such as wind, snowfall, snow storm, or rain during thaws. These enclosures are mostly pyramidal and are made of plastic film, tarpaulin, or rubberized fabric. Temperature inside them may be below zero in severe frosts but is always above the ambient air temperature, Such enclosures are often inflated or arch-framed type. The arch-framed enclosures are made of plastics or lightweight insulated sectional elements (panels) that have good heat-shielding properties. Heaters can be installed in such enclosures if necessary and so they can become heated enclosures.

Unheated enclosures are covered with dense windproof and waterproof materials to protect the work area against the ambient weather conditions.

The temperature conditions of concrete curing in such enclosures do not differ much from the ambient temperature. So the temperature conditions favorable for concrete hardening are provided inside unheated enclosures by heating concreted structures with any cost-effective and energy-efficient method. The presence of an enclosure reduces energy consumption for concrete heating and create conditions for continuous erection of reinforced concrete structures regardless of the environment.

Heated enclosures are most typical of this method for curing concrete in cast-in-place structures and the main requirement for their covering material is good heat-shielding properties. The most complex types of structures concreted in enclosures are cast-in-place reinforced concrete elements erected in sliding forms. So we describe this case below. Curing of concrete in such structures is limited by the pace of erection. Since temperature in an enclosure does not provide even normal temperature conditions for the hardening of concrete, we have to take additional measures to heat it to accelerate hardening and so to be able to raise the forms faster. There are two arrangements for heaters, including directly in sliding forms or on the floor of suspended scaffolds. Should there be forced delays in the operation, the zone of intensive concrete heating is moved to the lower area of the forms and to the adjacent part of the structure, from where the forms were already removed.

When heaters are installed directly in the formwork, concrete is placed, compacted, and starts to be heated with the temperature rising no faster than 20°C per hour. But the concrete cannot stay in the formwork for long because it must leave it with a strength no higher than 0.4 or 0.5 MPa. A higher strength increases adhesion of the concrete to the forms. So the concrete is heated mostly outside the formwork in an internal enclosure and an external enclosure which are located below the level of the platform where the concreting takes place. The temperature conditions in the enclosure below the work platform becomes then of a particular importance.

Special observations of the formation of the temperature field in enclosures and in concrete were made during construction of a multistory building with a cast-in-place core in Moscow.

The core 57.4 m high and measuring 15 × 12.6 in plan had three sections — two side sections 3 × 12.6 m in size and a central one of 9 × 12.6 m. The central section of the core was concreted in an enclosure that was not pyramidal. The enclosure was installed over the internal and external suspended scaffolds. The frame of the enclosure was a flexible system that had two degrees of freedom — it could be raised together with a system of jack frames and its suspended floor was movable horizontally to a certain degree. The height of the suspended enclosure was 3.2 m, the width of external suspended scaffolds was 1.2 m with an external cover made of plywood four mm thick. The internal enclosure was larger and the total heated volume of the enclosure was 816 m^3. The upper platform where concrete was placed in the forms was not covered. The study covered concreting at the heights from 13 to 21.5 m. The cast-in-place structure was heated in the enclosures with electric blowers installed on the floor of the external and internal suspended scaffolds. The stream of warm air was directed toward the concrete surface at an angle of 45° at 0.8 to one m from the floor of the scaffolds.

The temperature differences in the enclosures were great when the ambient air temperature was –6°C and the wind velocity three to six m/sec. It was 1.5 to two times lower in the external enclosure (between 6 and 16°C) than in the internal enclosure (from 10 and 20°C). The temperature of the concrete was approximately at the same level as that in the enclosures being 10 to 18°C.

When analyzing this concrete heating method we should note that it was far from perfect and did not ensure a fast strength development in concrete. There was also another drawback in the adopted method of heating. The streams of warm air directed at the concrete surface increased the loss of moisture from the material and thus could overdry the surface layer of the concrete in the structure. The volume being heated was not sealed tight, gaps between the boundaries of the cast-in-place structure and the suspended floor of the enclosure were one to eight cm wide because the floor was movable. So the air exchange rate in the movable enclosures was six within an hour. The adopted heating method proved to be quite energy consuming and, most important, did not enable the concrete to attain a strength of 70% of its design value when it emerged from the enclosure. The low concrete temperatures and slow strength development reduced the speed of the sliding forms and of the enclosure.

Is it possible to have a faster strength development in the concrete when reinforced concrete structures are cast in place in sliding forms that are raised at high speed? Yes, it is but other heating methods should be used for this. There are two better ways of doing it, including heating with flexible heating mats and the use of heating wires.

Heating mats are suspended in sliding forms as close as possible to concrete. When the temperature of the surface of a mat that faces the concrete is 60 to 80°C, the temperature on the surface of the latter is about 10°C lower because of the existing air gap between the mat and the structure since it is difficult to have the mat press tightly to the concrete. The external surface of the mat is insulated to reduce the heat loss to the environment. When the sliding forms are raised at three m/day, the concrete has a strength of at least 70% of the design value when it emerges from the enclosure. Power consumtion with this concrete heating method is several times less as compared with the convective heating by air blowers. When heated with heating mats, concrete loses much less moisture and never overdries.

The heating wire is installed on the reinforcement of the structure to be erected close to the concrete covers of the internal surface and of the external surface at a pitch of 10 cm. Once concrete is placed in a lift voltage is supplied to the wire and the heating starts. The voltage usually does not exceed 60 V for the safety reasons and the temperature on the wire surface is up to 80°C. The exposed surface is always covered with heat insulation mats that are suspended from the sliding forms. Heating wires consume less electric power than heating mats.

When structures are erected in sliding forms at low subzero temperatures (below –10°C), 2 to 4% of sodium nitrite by cement weight are better be added to the concrete mixture. It accelerates the hardening of concrete and further strength development when the enclosure is removed from the structure.

The above-described example of erecting a cast-in-place reinforced concrete structure in enclosures and sliding forms is the most complicated arrangement. In all other cases the concreting of structural members in enclosures (foundations, columns, floor slabs, etc.) simplifies the task of making the concrete harden faster. The most economical method for massive structures is electric preheating of concrete mixture followed by placing it immediately after the heating stops. Concrete can be heated by any electric heating method in a structure of low massiveness. Just like in any other case, the selection of a method depends on various technical, constructive, and economic factors.

11.3 AIR EXCHANGE IN ENCLOSURES

Air exchange is quite important for using enclosures in casting structures in place. If a lot of outside air penetrates an enclosure, energy consumtion for heating it goes up sharply. So to create the required heating conditions inside the enclosure it is necessary to prevent cold air from infiltrating through gaps at the level of the suspended floor over the perimeter where it is dajacent to the structure being erected.

T. M. Bochkareva, Cand.Sc. (Eng), proposed a sealing system for the concrete heating area and for the enclosure in general (Figure 11.1) by using hot air curtains to protect them against infiltration of outside air. The air are air flows oriented relative to the concrete surface. These continuous curtains are made by installing an air duct which is a single belt of pipes with holes whose diameter and spacing are found by calculation.

The air curtain on the level with the suspended floor ensures (in addition to sealing the enclosure) a positive pressure of hot air in the suspended cover. There is no infiltration of cold outside air in this case.

When concrete is heated is an enclosure with heaters, the maxium transfer of heat to the concrete from a heater can be provided by controlling the aerodynamics of air exchange in the closed air interlayer between the source of heat and the cast-in-place structure. It is well-known that the convective heating zone limited with the vertical surfaces of the structure being heated and of the continuous heater has two air flows. One is ascending along the hottest surface and the other is descending along the coldest surface. Friction between them depends on the width of the air

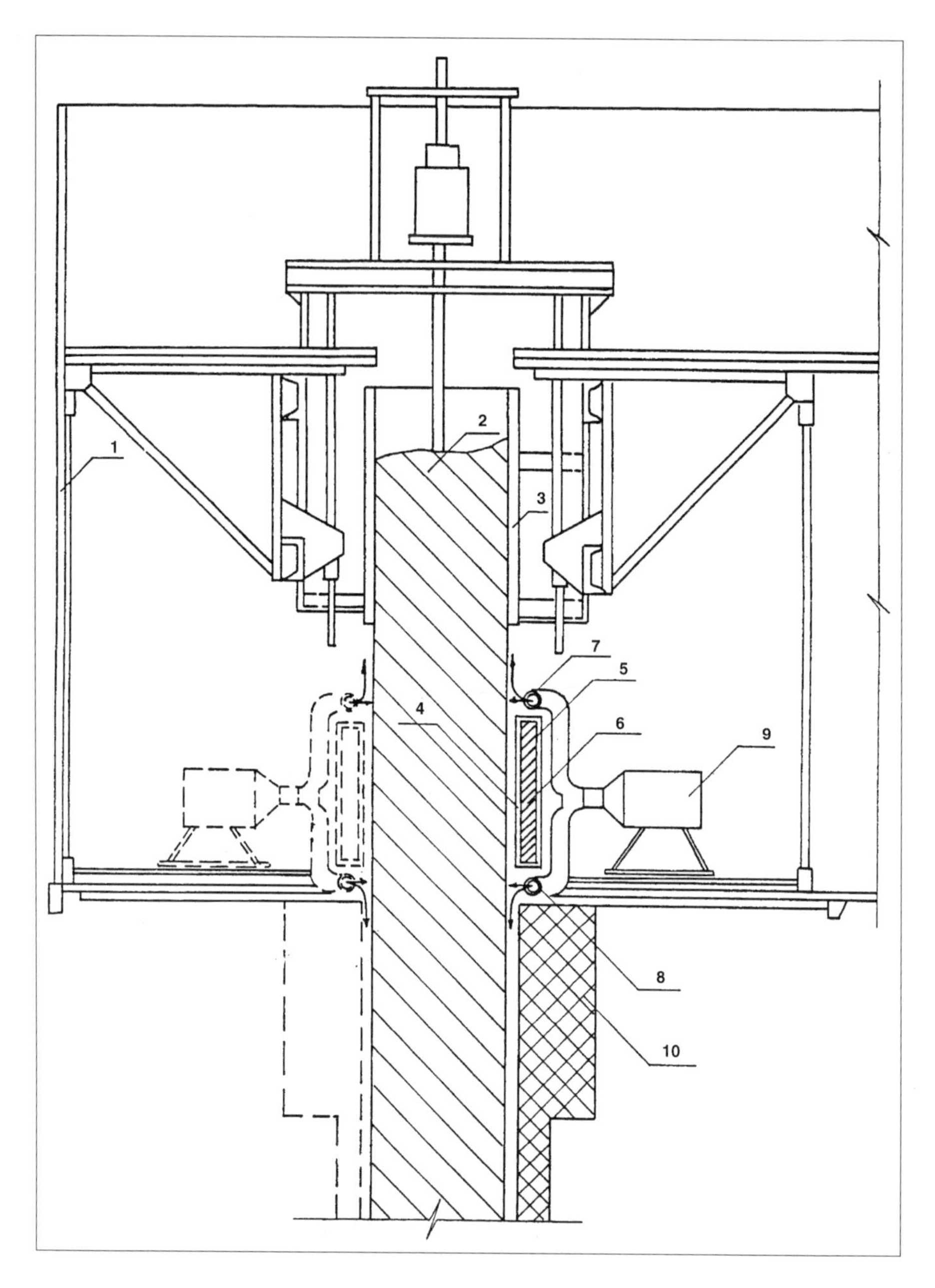

FIGURE 11.1 Heating a cast-in-place reinforced concrete structure erected in sliding forms in a movable round enclosure. 1 — cover of enclosure, 2 — concreted structure, 3 — sliding forms, 4 — electric heater panel, 5 — rigid box, 6 — heat insulation, 7, 8 — air ducts, 9 — blowers, 10 — suspended insulated cover.

interlayer. As it gets narrower, the probability increases that the ascending flow and the descending flow will merge into one mixed flow.

Two basic patterns of the dynamics of the stream are possible in the closed air interlayer between the heater and the surface of the structure — a continuous single loop of the ascending and descending air movement or several similar loops over the height of the air interlayer. A particualar pattern of air flows can be provided by creating an additional directed air stream to be included in the enclosure heating system in the form of an oriented air curtain. The air ducts are installed over the perimeter of the enclosure along the floor and near the top of the heaters designed for heating concrete in the enclosure. The continuous oriented air flows are thus created. It is preferable that the air curtain streams be directed at right angles to the heated surface of the cast-in-place structure in the upper and lower level of the system.

The air curtains may be made by pumping the preheated outside cold air into the ducts or by pumping the air from the heated space inside the enclosure with or without additional heating. When the heating system is on, a flow of hot air starts to rise along the surface of the heat source. It cannot leave the enclosure completely because it is stopped by the positive pressure created by the continuous air flow (air curtain) at the top. Under the constant pressure of the ascending hot air flow the latter fills the air interlayer over its height, which provides uniform heat transfer to the concrete from the heater.

The infiltration of cold air at the level of the suspended floor of the enclosure is prevented even when its doors are opened owing to the air curtain provided in the same plane with the floor of both internal and external suspended scaffolds and sealing them over the perimeter adjacent to the concreted structure.

It was found by calculations that the air curtains in a movable enclosure can reduce the air exchange rate there from six to 0.4^{-1}.

Heated air curtains energy expended for heating the enclosure and concrete and in general make the erection of cast-in-place reinforced concrete structures in enclosures more cost effective.

11.4 CONCRETING IN ENCLOSURES

Concreting in enclosures can be organized in different ways and depends on of the structure to be erected and the type of enclosure. If an enclosure is not heated, which is often the case with massive plain and reinforced concrete structures, concrete mixture is delivered from batching plants to the placing site in various vehicles. Then the enclosure has large openings with doors for the vehicles to enter and unload the concrete into bins or other equipment designed for feeding concrete to the structure. The structures being concreted in enclosures are protected against snow or ice formation and this is very convenient for the concreting operations. Placed concrete is cured by either the thermos method or under the conditions called "electric thermos" wherein concrete is heated to a required temperature whereupon the heating is stopped and the temperature is raised or maintained in the concrete due to heat of hydration.

But unheated enclosures can be used also for thin-walled reinforced concrete structures, particularly in regions with strong winds and heavy snofalls and snow storms. The heating of concrete in such enclosures needs less energy. The concrete can be heated by any method which is selected according to the type of structure, its reinforcement, size, etc.

Enclosures are also convenient to use to preheat concrete mixtures electrically in concreting structures of appropriate massiveness. The heat loss in this case is lower during heating than in the open air and the concrete cools down slower in the structure because of the absence of wind. Unheated types are good for erecting a reinforced concrete structure in the same enclosure where a foundation was installed for it. This arrangement is acceptable, of course, only for small one-story structures because to install a large and tall enclosure is a difficult engineering problem.

Concreting in heated enclosures is organized differently. Warm enclosures are usually installed to build tall structures and they are raised with a structure. Concrete is delivered to the enclosure along the interior shaft of the structure by lifts or other methods. This calls for special devices to transport the concrete directly into the structure, such as a telpher or a belt conveyor. If the height of a structure allows to deliver concrete by a crane or by a pump installed outside, an opening with a door is made in the enclosure to receive the concrete. Air curtains are convenient for this purpose because they reduce the loss of warm air from the enclosure to the environment. But even in this enclosure the hardening of concrete can be accelerated, as was mentioned above. only by heating it with various methods. The air temperature inside the enclosure is not sufficient to make the concrete harden rapidly and thus accelerate a construction project in general.

Great care should be taken to ensure fire safety during the use of any enclosure, particularly at a great height. All wooden elements of the enclosure and covering materials (tarpaulin and the like) should be impregnated with fireproof compounds. The electric wiring for lighting, concrete heating, voltage supply to the electric equipment (vibrators, electric tools, etc.) should be carefully insulated, placed where they cannot be damaged, and inspected regularly by an electrician. A fast escape route should be provided for the personnel through the interior shaft in case of fire.

12 Some Problems of Economics of Winter Concreting

Reinforced concrete structural members are main elements in construction of most buildings and structures, particularly for industrial purposes, by volume of the material they use. They are expensive and to cast them in place in winter is more difficult and costly than in the summertime. Construction projects of this type always involve feasibility studies that can be divided into two stages.

The first stage is evaluation of the expediency of concreting in winter in general. It involves estimation of additional costs as compared with the same work done in the warm season. The data thus obtained serve as a basis for making the decision whether to start the concreting later in the year when the temperature conditions become more favorable or to go ahead with the construction despite higher costs. The work schedule for the construction and the commissioning of the project are predominant considerations at this stage.

When a large industrial complex is erected or several projects are built at the same time, the work can always be organized so that the concreting schedule is moved to the warm season. But if it cannot be done, we have to decide how we will erect reinforced concrete structures in the frost.

The second stage involves selecting winter concreting methods such that the additional money and labor costs are minimized not to the detriment, of course, to the quality and durability of the structures to be erected.

The first stage of the economic calculation determines all the costs involved in concreting in the frost. They include (1) the cost of the equipment and heat required to warm up concrete constituents, (2) the cost of transportation of concrete mixtures at low temperatures so that they do not freeze, (3) energy consumption for preparation of the formwork, (4) curing of concrete until it attains the required strength, (5) the cost of the equipment and of heating of rooms for workers, and (6) all other expenditures that depend on the ambient temperature and conditions at the site. It is the analysis of the additional costs involved in winter concreting taking into account the work schedule, the duration of the winter season and some other factors that makes it possible to decide whether it makes sense to concrete in winter.

If the decision is made in favor of winter concreting, the problem at the second stage of the economic analysis is reduced to the selection of the most efficient organization of the work to minimize the costs in excess of those involved in concreting in summer.

When the volume of concrete to be placed is large, it is better to install the batching plant at the construction site to avoid high transportation costs. Energy is d most of all if the method of accelerating the hardening of concrete was chosen correctly. The methods used to cure placed concrete may be different for different structures according to technological and other considerations because each method has its most efficient application.

Thus, during the construction of the Church of Christ the Savior in Moscow, which was carried out the whole year round and on a very tight schedule, thermos curing in an insulated formwork was used to accelerate concrete hardening in massive pillars. When heavily reinforced floors, cupolas, and other elements 10 to 30 cm thick were erected, concrete was heated with heating wires. 3 to 4% of sodium nitrite by dry weight of cement were added in mixing to protect the concrete mixture against freezing during transport and placing.

During the construction of the Manege Square Complex in Moscow, thermos curing for massive structures was supplemented by heating wires used for heating concrete in heavily reinforced columns and floors, electrode heating for erecting walls and some floors, and local enclosures for heating concrete poured into the bases of steel columns. Each of these methods was the most convenient and cost saving for curing concrete in particular structures and ensured their high quality. In addition to the thermal treatment, the sodium nitrite admixture (3 or 4% by cement weight), microsilica, or a superplasticizer had to be added to concrete mixtures according to the type of structure . This differentiation minimized the additional costs as compared with the warm season.

There is no universal recipe for concreting in the frost. Approaches to winter concreting may be different in each particular case and in each country according to the type of structure to be erected, local conditions, and capabilities of a building contractor. They are dictated by the duration of the winter period and by temperature levels, by the cost of electricity, of cement, of heat insulating materials, of chemical admixtures, etc.

Under any construction conditions and in any country the main criterion for casting concrete structures in place in winter should be their high quality, reliability, and durability. And the economic analysis should minimize the costs involved in achieving this goal.

Appendices

TABLE 1
Specific power required to heat concrete with strip electrodes arranged on one side (the thickness of the structure is 10 cm) in kW/m³

Voltage in V	Mean specific resistance in ohm.cm	Power in kW/m³							
		Distance between centers of electrodes in cm							
		10		20		30		40	
		Width of electrodes in cm							
		2	5	2	5	2	5	2	5
51	200	—	—	17.27/14.93	22.49/20.38	9.20/8.09	11.14/10.34	5.64/5.12	6.65/6.26
	800	11.76/9.69	17.19/14.84	4.32/3.73	5.62/5.09	2.28/2.02	2.79/2.59	1.41/1.28	1.66/1.57
	1600	4.88/4.84	8.59/7.42	2.16/1.87	2.81/2.54	1.14/1.01	1.39/1.29	0.71/0.64	0.83/0.78
60	200	—	—	23.91/20.66	31.12/28.20	12.59/11.20	15.42/14.32	7.81/7.08	9.20/8.67
	800	16.28/13.41	23.79/20.54	5.98/6.16	7.78/7.05	3.15/2.80	3.86/3.58	1.95/1.77	2.30/2.17
	1600	8.14/6.70	11.89/10.27	2.99/2.58	3.89/3.53	1.57/1.40	1.93/1.79	0.98/0.89	1.15/1.08
70	200	—	—	—	—	17.14/15.25	20.99/19.49	10.63/9.64	12.52/11.80
	800	22.16/18.25	32.38/27.96	8.14/7.03	10.59/9.60	4.28/3.81	5.25/4.87	2.66/2.41	3.13/2.95
	1600	11.08/9.13	16.19/13.98	4.07/3.52	5.30/4.80	2.14/1.91	2.62/2.44	1.33/1.21	1.57/1.48
87	200	—	—	—	—	—	—	16.41/14.89	19.34/18.23
	800	—	—	12.57/10.86	16.36/14.82	6.62/5.89	8.10/7.53	4.10/3.72	4.84/4.56
	1600	17.11/14.09	25.01/21.59	6.28/5.43	8.18/7.41	3.31/2.94	4.05/3.76	2.05/1.86	2.42/2.28
106	200	—	—	—	—	—	—	24.37/22.10	28.71/27.06
	800	—	—	18.66/16.12	24.29/22.01	9.83/8.74	12.03/11.17	6.09/5.52	7.18/6.36
	1600	25.41/20.92	37.12/32.05	9.33/8.06	12.14/11.00	4.91/4.37	6.02/5.59	3.06/2.76	3.59/3.38

TABLE 1 (continued)
Specific power required to heat concrete with strip electrodes arranged on one side (the thickness of the structure is 10 cm) in kW/m³

Voltage in V	Mean specific resistance in ohm.cm	Power in kW/m³							
		Distance between centers of electrodes in cm							
		10		20		30		40	
		Width of electrodes in cm							
		2	5	2	5	2	5	2	5
127	200	—	—	—	—	—	—	—	—
	800	—	—	—	—	14.11/12.55	17.27/16.03	8.74/7.93	10.30/9.71
	1600	—	—	13.39/11.57	17.43/15.80	7.05/6.27	8.64/8.02	4.37/3.97	5.15/4.86

Note: The numerator shows power for three-phase current and the denominator for single-phase current.

TABLE 2
Specific power required to heat concrete with strip electrodes arranged on one side (the thickness of the structure is 20 cm) in kW/m³

Voltage in V	Mean specific resistance in ohm.cm	Power in kW/m³							
		Distance between centers of electrodes in cm							
		10		20		30		40	
		Width of electrodes in cm							
		2	5	2	5	2	5	2	5
51	200	—	—	9.49/7.68	12.73/10.59	5.52/4.57	7.09/6.07	3.67/3.11	4.57/3.99
	800	5.55/4.35	7.90/6.32	2.37/1.92	3.18/2.65	1.38/1.14	1.77/1.52	0.92/0.75	1.14/1.00
	1600	2.78/2.17	3.95/3.16	1.19/0.96	1.59/1.32	0.69/0.57	0.89/0.76	0.46/0.38	0.57/0.50
60	200	—	—	13.13/10.62	17.62/14.66	7.63/6.33	9.81/8.40	5.08/4.30	6.33/5.53
	800	7.69/6.02	10.94/8.74	3.28/2.66	4.40/3.66	1.91/1.58	2.45/2.10	1.27/1.07	1.58/1.38
	1600	3.84/3.01	5.47/4.37	1.64/1.33	2.20/1.83	0.95/0.79	1.23/1.05	0.64/0.54	0.79/0.69
70	200	—	—	17.88/14.46	23.98/19.95	10.39/8.62	13.36/11.43	6.91/5.85	8.62/7.53
	800	10.46/8.92	14.89/11.90	4.47/3.62	6.00/4.00	2.60/—	3.34/—	1.73/—	2.15/—

TABLE 2 (continued)
Specific power required to heat concrete with strip electrodes arranged on one side (the thickness of the structure is 20 cm) in kW/m³

Voltage in V	Mean specific resistance in ohm.cm	Power in kW/m³							
		Distance between centers of electrodes in cm							
		10		20		30		40	
		Width of electrodes in cm							
		2	5	2	5	2	5	2	5
	1600	5.23/4.10	7.44/5.95	2.24/1.81	3.00/2.49	1.30/1.08	1.67/1.43	0.87/0.73	1.08/0.94
87	200	—	—	—	—	16.05/13.31	20.63/17.65	10.68/9.04	13.31/11.62
	800	16.16/12.65	22.99/18.38	6.90/5.58	4.01/7.70	5.16/3.33	2.67/4.41	2.67/2.26	3.33/2.97
	1600	8.08/6.33	11.50/9.19	3.45/2.79	4.63/3.85	2.01/1.67	2.58/2.21	1.34/1.13	1.66/1.45
106	200	—	—	—	—	22.83/19.73	30.63/26.21	15.85/13.42	19.76/17.25
	800	23.99/18.78	34.14/27.29	10.25/8.29	13.75/11.44	9.96/4.94	7.66/5.55	3.96/3.35	4.94/4.31
	1600	12.00/9.39	17.07/13.65	5.12/4.15	6.87/5.72	2.98/2.46	3.83/3.28	1.98/1.68	2.47/2.16
127	200	—	—	—	—	—	—	—	—
	800	—	—	14.70/11.90	17.73/16.42	8.55/7.07	10.99/9.40	5.69/4.81	7.09/6.19
	1600	17.22/13.48	24.50/19.59	7.36/5.95	9.87/8.21	4.28/3.56	5.50/4.70	2.85/2.41	3.55/3.10

Note: The numerator shows power for three-phase current and the denominator for single-phase current.

TABLE 3
Specific power required to heat concrete with strip electrodes arranged on one side (the thickness of the structure is 30 cm) in kW/m³

Voltage in V	Mean specific resistance in ohm.cm	Power in kW/m³							
		Distance between centers of electrodes in cm							
		10		20		30		40	
		Width of electrodes in cm							
		2	5	2	5	2	5	2	5
51	200	13.80/10.60	19.1/14.8	6.22/4.90	8.30/6.65	3.79/3.04	4.92/4.03	2.62/2.13	3.34/2.78

TABLE 3 (continued)
Specific power required to heat concrete with strip electrodes arranged on one side (the thickness of the structure is 30 cm) in kW/m³

Voltage in V	Mean specific resistance in ohm.cm	Power in kW/m³							
		Distance between centers of electrodes in cm							
		10		20		30		40	
		Width of electrodes in cm							
		2	5	2	5	2	5	2	5
	800	3.45/2.65	4.78/3.70	1.56/1.23	2.08/1.66	—	—	—	—
	1600	1.73/1.33	2.39/1.85	—	—	—	—	—	—
60	200	19.10/14.62	2.62/20.20	8.58/6.76	11.42/9.17	5.24/4.20	6.80/5.56	3.62/2.94	4.61/3.84
	800	4.77/3.66	6.60/5.11	2.15/1.70	2.87/2.29	—	—	—	—
	1600	2.39/1.84	3.30/2.55	—	—	—	—	—	—
70	200	26.00/19.9	35.91/27.90	11.68/9.22	15.60/12.49	7.12/5.72	9.25/7.56	4.92/4.01	6.27/5.22
	800	6.48/4.98	9.00/6.95	2.93/2.32	3.91/—	—	2.31/1.90	—	—
	1600	3.25/2.49	4.50/3.44	—	—	—	—	—	—
87	200	—	—	18.12/14.27	24.17/19.34	11.02/8.85	14.29/11.17	7.63/6.19	9.72/8.10
	800	10.05/7.72	13.92/10.74	4.54/3.58	6.05/4.83	2.76/2.21	3.57/2.94	—	2.44/2.02
	1600	5.04/3.87	6.95/5.38	2.27/1.79	3.02/2.41	—	—	—	—
106	200	—	—	26.85/21.19	35.84/28.73	16.33/13.10	21.20/17.40	11.30/9.20	14.40/11.98
	800	14.96/11.50	20.60/15.93	6.65/5.32	8.98/7.16	4.10/3.28	5.32/4.36	2.84/2.28	3.62/3.02
	1600	7.48/6.55	10.30/7.99	3.37/2.64	4.49/3.58	2.05/1.64	2.66/2.18	—	—
127	200	—	—	38.55/30.38	51.50/41.21	23.48/18.82	30.43/24.98	16.24/13.20	20.62/17.23
	800	21.4/16.43	29.60/22.94	9.68/7.56	12.88/10.20	6.84/4.82	7.64/6.25	4.09/3.28	5.21/4.34
	1600	10.70/8.22	14.80/11.47	4.84/3.78	6.44/5.25	2.92/2.36	3.82/3.12	2.05/1.64	2.61/2.17

Note: The numerator shows power for three-phase current and the denominator for single-phase current.

TABLE 4
Specific power required to heat concrete with strip electrodes arranged on one side (the thickness of the structure is 40 cm) in kW/m³

Voltage in V	Mean specific resistance in ohm.cm	Power in kW/m³							
		Distance between centers of electrodes in cm							
		10		20		30		40	
		Width of electrodes in cm							
		2	5	2	5	2	5	2	5
51	200	9.75/7.44	10.50/10.21	4.53/3.52	6.00/4.79	2.82/2.22	3.66/2.92	1.99/1.58	2.53/2.06
	800	2.44/1.86	2.68/2.55	—	—	—	—	—	—
	1600	—	—	—	—	—	—	—	—
60	200	13.44/10.24	14.50/14.05	6.25/4.85	8.27/6.61	3.89/3.06	5.05/4.03	2.74/2.18	3.49/2.84
	800	3.36/2.56	3.62/3.51	—	2.07/1.65	—	—	—	—
	1600	—	—	—	—	—	—	—	—
70	200	18.32/14.00	19.73/19.18	8.52/6.61	11.28/9.02	5.30/4.17	6.88/5.48	3.74/2.97	4.76/3.88
	800	4.57/3.50	4.93/4.79	2.13/1.65	2.82/2.25	—	—	—	—
	1600	2.29/1.75	2.46/2.39	—	—	—	—	—	—
87	200	28.39/21.64	30.28/29.64	13.18/10.22	17.48/13.97	8.20/6.46	10.62/8.50	5.81/4.60	7.36/6.00
	800	7.10/5.40	7.57/7.40	3.30/2.56	4.37/3.51	2.05/1.62	2.65/2.13	—	—
	1600	3.55/2.72	3.79/3.70	—	2.18/1.75	—	—	—	—
106	200	—	—	19.60/15.21	25.93/20.71	12.18/9.60	15.79/12.61	8.60/6.83	10.93/8.91
	800	10.50/8.00	11.31/10.95	4.75/3.78	6.30/4.04	3.05/2.43	3.95/3.14	2.15/1.72	2.75/2.24
	1600	5.25/4.00	5.55/5.37	2.45/1.89	3.22/2.58	—	—	—	—
127	200	—	—	28.10/21.83	37.20/29.79	17.50/13.79	22.90/18.10	12.13/9.79	15.17/12.78
	800	15.70/11.51	16.28/15.80	7.03/5.46	9.30/7.45	4.37/3.45	5.73/4.52	3.08/2.48	3.92/3.20
	1600	7.56/5.76	8.14/7.90	3.51/2.73	4.75/3.73	2.18/1.73	2.86/2.26	—	—

Note: The numerator shows power for three-phase current and the denominator for single-phase current.

References

Abramov, V. S., Danilov, N. N., and Krasnovsky, B. M., *Elektrotermoobrabotka betona* (The Electric Thermal Treatment of Concrete), Moscow: Institute of Civil Engineering, 1975.

Arbeniev, A. S., *Winter Concreting With Electric Heating of Concrete Mixtures*, Promyshlennoe Stroitelstvo, 1962, No. 9.

Cold Weather Concreting, American Concrete Institute, Committee 306, ACI 3006, R-78, 1978.

Concrete Materials and Methods of Concrete Construction, National Standard of Canada CAN3-A23. 1-M77, Canadian Standards Association, 1977.

Eroshkin, V. N., Krylov, B. A., Pazyuk, Yu. V., and Kinilikova, G. T., *Vozvedenie konstruktsii iz monolitnogo betona v zimnikh usloviyakh* (Casting Structures in Place under Winter Conditions), Frunze: NIINTI, 1989.

Fagerlund, G., *Frost Resistance of High Performance Concrete — Some Theoretical Considerations*, Lund: Div. of Building Materials, Lund Institute of Technology, Report TVBM-3056, 1994.

Fagerlund, G., *Influence of Slag Cements on the Frost Resistance of Green Concrete*, The Third International RILEM Symposium on Winter Concreting, Finland, ESPOO, 1985.

Fagerlund, G., *Moisture Uptake and Service Life of Concrete Exposed to Frost*, Proceedings of the International Conference on Concrete under Severe Conditions, Supporo, 1995.

Fagerlund, G., *On the Service Life of Concrete Exposed to Frost Action*, Lund: Div. of Building Materials, Lund Institute of Technology, Report TVBM-7054, 1993.

Gnyrya, A.I., *Tekhnologiya betonnykh rabot v zimnikh usloviyakh* (Concreting Procedures under Winter Conditions), Tomsk: Tomsk University, 1984.

Ivanov, F. M., *Testing Methods and Frost Resistance Standardization of Concrete*, Proceedings of the International Conference on Concrete under Severe Conditions, Supporo, 1995.

Kilpi, E. and Kukko, H., *Properties of Hot Concrete and Its Use in Winter Concreting*, Oslo: Nordic Concrete Federation, Publication No. 1, 1982.

Kilpi, E. and Sarja, A., *Builder's Guide to Safe Winter Concreting*, Research Notes 62, Technical Research Centre of Finland, ESPOO, 1981.

Kivekas, L. and Leivo, M., *Research on Use of Antifreeze Admixtures in Finland*, The Third International RILEM Symposium on Winter Concreting, Finland, ESPOO, 1985.

Kivekas, L., Huovinen, S., and Leivo, M., *Concrete under Arctic Conditions*, Research Report 343, VTT of Finland, ESPOO, 1985.

Komissarov, L.A., *Opyt primeneniya elektroprogreva betona i zhelezobetona* (The Experience in Electric Heating Plain and Reinforced Concrete), Moscow: Gosstroyizdat, 1956.

Krylov, B. A. and Li, A. I., *Forsirovannyi razogrev betona* (Accelerated Concrete Heating), Moscow: Stroyizdat, 1975.

Krylov, B. A. and Pizhov, A. I., *Teplovaya obrabotka betona v greyushchei opalubke s setchatymi elektronagrevatelyami* (The Thermal Treatment of Concrete in Heating Forms with Mesh Heaters), Moscow: Stroyizdat, 1975.

Krylov, B. A. and Zvezdov, A. I., *Temperature Influence on Concrete Structures and Its Hardening, Proceedings of the International Conference on Concrete under Severe Conditions*, Supporo, 1995.

Krylov, B. A., *Effektivnoe resursosberezhenie* (Effective Resource Saving), Moscow: Znanie, 1989.

Krylov, B. A., Mukha, V. I., and Abakumov, Yu. N., *Zadelka stykov sbornykh zhelezobetonnykh konstruktsii v zimnee vremya* (Sealing Joints between Precast Reinforced Concrete Units in Winter Time), Moscow: Stroyizdat, 1966.

Kukko, H., *Use of Heated Fresh Concrete*, The Third International RILEM Symposium on Winter Concreting, Finland, ESPOO, 1985.

Kykko, H., *Frost Effects on High Strength Concrete Without Air Entrainment*, Proceedings of the International Conference on Concrete under Severe Conditions, Supporo, 1995.

Manual of Concrete Practice. Part 2: Cold Weather Concreting, American Concrete Institute, 1985.

Mironov, S. A., *Teoriya i metody zimnego betonirovaniya* (The Theory and Methods of Winter Concreting), Moscow: Gosstroyizdat, 1974.

Mironov, S. A., Vegener, R. V., and Semensky, K. P., *Electric Heating of Concrete*, ONTI, 1938.

Möller, G, *Early Freezing of Concrete*, Applied Studies No. 5, Stockholm: Swedish Cement and Concrete Research Institute, 1962.

Nichols, M. R. and Christensen, G. S., *Concrete Mixture Proportioning Techniques for Remote Alaskan Locations*, Proceedings of the International Conference on Concrete under Severe Conditions, Supporo, 1995.

Penttala, V., *Heating of Concrete by Infra-Red Radiation*, The Third International RILEM Symposium on Winter Concreting, Finland, ESPOO, 1985.

Pigeon, M. and Regourd, M., *The Effects of Freeze-Thaw Cycles on the Microstructure of Hydration Particles*, Durability of Building Materials, 1986, Vol. 4, no. 1.

Podvalny, A. M., *Mechanism of Deterioration of Concrete under Severe Environment*, Proceedings of the International Conference on Concrete under Severe Conditions, Supporo, 1995.

Powers, T. C., *A Working Hypothesis for Further Studies of Frost Resistance of Concrete*, ACI Journal, 1945, Vol. 41.

Powers, T. C., *Resistance of Concrete to Frost at Early Ages*, Proceedings of the RILEM Symposium on, Copenhagen, 1956.

Ronin, V. and Jonasson, E., *Energetically Modified Cement for High Performance Concrete in Winter Concreting*, Proceedings of the International Conference on Concrete under Severe Conditions, Supporo, 1995.

Rukovodstvo po elekrotermoobrabotke betona (Guidelines for the Electric Thermal Treatment of Concrete), Moscow: NIIZhB, Stroyizdat, 1974.

Rukovodstvo po zimnemu betonirovaniyu s primeneniem metoda termosa (Guidelines for Winter Concreting With the Thermos Method), Moscow: NIIZhB, Stroyizdat, 1975.

Sakai, K., *Cold Weather Concreting in Hokkaido* (in Japanese), Concrete Journal, 1991, Vol. 29, no. 6.

Sakai, K., Watanabe, H., and Hamabek, *Antifreeze Admixture Development in Japan*, Concrete International, 1991.

Sarja, A. and Punakkalio, E., *Development of Quality Assurance for Winter Concreting*, The Third International RILEM Symposium on Winter Concreting, Finland, ESPOO, 1985.

Sarkar, S. L. and Malhotra, V. M., *Microstructure Durability of Concretes Exposed to Arctic Conditions*, Proceedings of the International Conference on Concrete under Severe Conditions, Supporo, 1995.

Scanlon, J. M., *Cold Weather Concreting in the U.S.A.*, The Third International RILEM Symposium on Winter Concreting, Finland, ESPOO, 1985.

Standard Specification for Cold Weather Concreting, American Concrete Institute, Committee 306, Draft No. 10, ACI, 1985.

Stark, J. and Ludwig, H.-M., *Effects of Low Temperature and Freeze-Thaw Cycles on the Stability of Hydration Products*, The 9th International Congress on the Chemistry of Cement, New Delhi, 1992, Vol. 4.

Verbeck, G. and Landgren, R., *Influence of Physical Characteristics of Aggregates on Frost Resistance of Concrete*, Proceedings of the American Society for Testing and Materials, 1960, Vol. 60.

Wang, K, Monteiro, P., Rubinsky, B., and Arav, A., *Microscope Study of Ice Propagation in Concrete*, ACI Journal, 1995.

Yamashita, H., Kita, T., Sakai, K., and Takahashi, J, *Frost Damage of Concrete Structures in Hokkaido*, Proceedings of the International Conference on Concrete under Severe Conditions, Supporo, 1995.

Yoshiro Koh, *Cold Weather Concreting in Japan*, The Third International RILEM Symposium on Winter Concreting, Finland, ESPOO, 1985.

Index

D

E

F

G

T